Signal Conditioning & PC-Based Data Acquisition Handbook

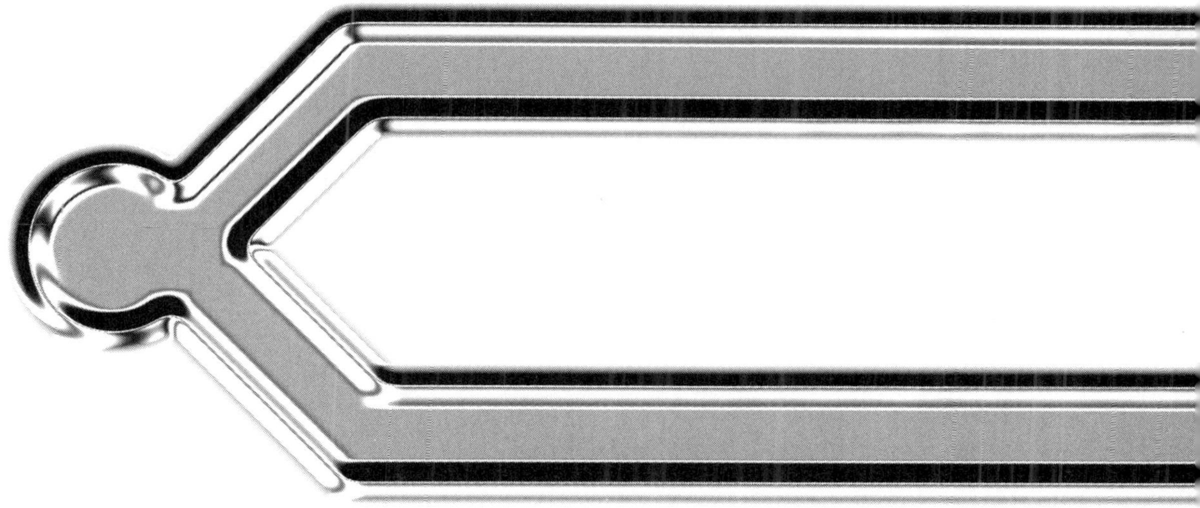

Edited By John R. Gyorki

A Reference For Analog & Digital Signal Conditioners & PC-Based Data Acquisition Systems

Third Edition

Copyright Notice

Published 2004

Printed in the United States of America

ISBN (Paperback): 0-9656789-3-8

Table of Contents

Preface to the Third Edition

The third edition of this handbook has been totally revised to include new chapters on Electrical Measurements, Vibration and Sound, Displacement and Position Sensing, and Transducer Electronic Data Sheets (TEDS). It also includes several new subjects and expands on selected items including Fundamental Signal Conditioning.

All chapters have been enhanced to address more practical applications than theoretical measurement issues. They cover a major topic with sufficient detail to help readers understand the basic principles of sensor operation and the need for careful system interconnections. The handbook also discusses key issues concerning the data acquisition system's multiplexing and signal conditioning circuits, and analog-to-digital converters. These three functions establish the overall accuracy, resolution, speed, and sensitivity of data acquisition systems and determine how well the systems perform.

Data acquisition systems measure, store, display, and analyze information collected from a variety of devices. Most measurements require a transducer or a sensor, a device that converts a measurable physical quantity into an electrical signal. Examples include temperature, strain, acceleration, pressure, vibration, and sound. Yet others are humidity, flow, level, velocity, charge, pH, and chemical composition.

Sensors come in numerous shapes, sizes, and specifications. They connect between the measured physical device and the signal conditioner's input. Most sensors are purchased off-the-shelf, but in some cases, they are custom made specifically for a particular measurement requirement. Regardless of input, however, the output signal is usually a voltage, current, charge, or resistance and all can be conditioned and handled equally well. Manufacturers frequently provide specifications, application notes, and principles of operation for their specific sensor to help users apply the device in the most efficient way.

Signal conditioners accept sensor output signals and convert them into a form that the data acquisition system can manipulate. Signal conditioners typically amplify, filter, isolate, and linearize these signals. They also convert current to voltage and voltage to frequency, provide other functions such as simultaneous sample and hold (SS&H), and supply a bias voltage or signal excitation for certain transducers. They may come with single-ended inputs or differential inputs for improving signal-to-noise ratios. The output of the signal conditioner, in turn, connects to the input of an analog-to-digital converter (ADC) embedded within the data acquisition system. Finally, the ADC converts the conditioned analog signal to a digital signal that can be transferred out of the data acquisition system to a computer for processing, graphing, and storing.

Introduction to Data Acquisition and Signal Conditioning

Chapter 1 discusses signals, sensors, and signal-conditioning techniques and how they relate to data acquisition system fundamentals. It also covers personal computers and how laptop or notebook computers work with data acquisition systems.

Analog-to-Digital conversion.

Chapter 2 discusses four basic ADC types, including their accuracy and resolution. Also covered are topics such as ADC output averaging, discrete sampling, input and source impedance, and differential voltage measurements. Yet others include simultaneous sample and hold methods, selectable input ranges, aliasing, digital filtering, and Fourier Transforms.

Multiplexing

Chapter 3 covers the fundamental principles of multiplexing and their benefits and economies.

Electrical Measurements

Chapter 4 and Chapters 6 through 9 discuss basic electrical measurements, the characteristics of various sensors, how to use sensors to measure the most common types of electrical and physical quantities, and the signals that they represent. Topics include voltage, current, resistance, charge, temperature, strain, position, velocity, acceleration, and sound. The sensors that produce them include thermocouples, RTDs, thermistors, strain gages, accelerometers, and linear and rotational displacement sensors.

Fundamental Signal Conditioning

Chapters 5 and 11 discuss the most widely used techniques for analog, digital, and pulse-train signal conditioning. They comprise operational, differential, and high-gain amplifiers for filtering, attenuation, isolation, linearization, and circuit protection for analog signals. They also cover topics on digital I/O interfacing, frequency measurements, and pattern generation for digital signals.

Noise Reduction and Isolation

Chapter 10 is dedicated to simplifying the somewhat difficult topic of electrical noise interference, using the best shielding and grounding techniques, and identifying the major sources of crosstalk. It also discusses how to select the proper amplifiers and sensors, as well as how to use certain isolation and wiring techniques to minimize or eliminate significant noise in data-acquisition systems.

Transducer Electronic Data Sheets

Chapter 12 covers Transducer Electronic Data Sheets. T.E.D.S. is a class of so-called smart sensors that contain an onboard memory chip. The chip stores information regarding transducer calibration, manufacturer information, and many other data.

Contact IOtech

Contact the IOtech Applications Department for more detailed descriptions of the signal conditioning schemes discussed, product specification sheets, and application notes. Also, e-mail your comments regarding the handbook's content and accuracy to **productsupport@iotech.com**. Find application notes, product descriptions, instruction manuals, and other information at IOtech's Web site, **www.iotech.com**.

Chapter 1

INTRODUCTION TO DATA ACQUISITION AND SIGNAL CONDITIONING

SIGNALS, SENSORS, AND SIGNAL CONDITIONING

All industrial processing systems, factories, machinery, test facilities, and vehicles consist of hardware components and computer software whose behavior follow the laws of physics as we understand them. These systems contain thousands of mechanical and electrical phenomena that are continuously changing; they are not steady state. The measurable quantities that represent the characteristics of all systems are called variables. The proper functioning of a particular system depends on certain events in time and the parameters of these variables. Often, we are interested in the location, magnitude, and speed of the variables, and we use instruments to measure them. We assign the variables units of measure such as volts, pounds, and miles per hour, to name a few.

Most variables must be measured with a device that converts the phenomena into a form that a human can perceive such as a visual display, a transducer for sound, or vibrations to stimulate physical sensations. The conversion devices are called transducers or sensors, and they translate the physical phenomena to electrical signals (or vice versa) to be measured with electronic instruments. These instruments have traditionally been ammeters, voltmeters, and various other gages, and the variables can be observed in real time. But an increasing need to record and preserve these phenomena and analyze them at a later time forced engineers to develop data recorders and data acquisition systems.

Variables may be classified in many ways, but generally, most experts prefer two classifications: by characteristic and by type of measurement signal. Variables classified by characteristic include thermal, radiation, force, rate, quantity, time, geometric, physical properties, chemical composition, and electrical. Those classified by measurement signal include motion, force, electrical, and time-modulated. Measurement signals for variables often are hard to differentiate from the measuring system. Four factors require close consideration for measurement signals and systems: the types of transducers available for converting variables to measurement signals, transmission characteristics, data acquisition system input matching, and transducers available to convert from one type of measurement signal to another measurement signal.

DATA ACQUISITION SYSTEMS

Data acquisition systems have evolved over time from electromechanical recorders containing typically from one to four channels to all-electronic systems capable of measuring hundreds of variables simultaneously. Early systems used paper charts and rolls or magnetic tape to permanently record the signals, but since the advent of computers, particularly personal computers, the amount of data and the speed with which they could be collected increased dramatically. However, many of the classical data-collection systems still exist and are used regularly.

PC–BASED DATA-ACQUISITION EQUIPMENT

Early, expensive mainframe computers were used extensively for gathering multiple channels of data, primarily in large industrial or scientific applications. They were seldom used in small projects because of their relatively high cost. But the introduction of small rack-mounted minicomputers that developed in the 1960's and later desktop personal-type computers that housed microprocessors and proliferated in the 1970's justified their use for smaller projects. Soon, data acquisition plug-in cards (as well as hundreds of other types of plug-in cards) for these small computers were a common means to collect and record data of all types.

Plug-in cards for computers did not always perform to the user's expectations, however. Internal noise from rotating devices such as drives and electromagnetic and electrostatic noise from the computer's internal bus structure often interfered with the measured variable, particularly in data acquisition cards. Isolation and shielding have helped to solve the problem in most cases, but many data acquisition manufacturers also provide signal conditioning and signal processing circuits in small, stand-alone, shielded enclosures. The separate box provides isolation by distance, expansion for hundreds of channels, and portability with laptop computers that desktop personal computers with plug-in cards don't possess.

All PC-based data acquisition systems will record extremely accurate, repeatable, reliable, and error-free data provided they are connected and operated according to the manufacturer's recommended practices. These practices include selecting the correct sensors for the application, the proper wire and shielded cable; capturing the signals in proper magnitude, range, and frequency; and paying close attention to grounding and shielding – particularly eliminating ground loops. Additional items include choosing the correct impedance and using doubled-ended (differential) inputs instead of single-ended where possible. The environment should also be considered, especially for extremes of ambient temperature, shock, and vibration. And herein lies the major goal of this publication – to inform users of the most needed recommended practices based upon a fundamental knowledge of the internal workings of data acquisition system instrumentation.

Chapter 2

ANALOG-TO-DIGITAL CONVERSION

ADC TYPES

Analog-to-Digital Converters (ADCs) transform an analog voltage to a binary number (a series of 1's and 0's), and then eventually to a digital number (base 10) for reading on a meter, monitor, or chart. The number of binary digits (bits) that represents the digital number determines the ADC resolution. However, the digital number is only an approximation of the true value of the analog voltage at a particular instant because the voltage can only be represented (digitally) in discrete steps. How closely the digital number approximates the analog value also depends on the ADC resolution.

A mathematical relationship conveniently shows how the number of bits an ADC handles determines its specific theoretical resolution: An n-bit ADC has a resolution of one part in 2^n. For example, a 12-bit ADC has a resolution of one part in 4,096, where 2^{12} = 4,096. Thus, a 12-bit ADC with a maximum input of 10 Vdc can resolve the measurement into 10 Vdc/4096 = 0.00244 Vdc = 2.44 mV. Similarly, for the same 0 to 10-Vdc range, a 16-bit ADC resolution is $10/2^{16}$ = 10/65,536 = 0.153 mV. The resolution is usually specified with respect to the full-range reading of the ADC, not with respect to the measured value at any particular instant.

Successive-Approximation ADCs

A successive-approximation converter, Figure 2.01, is composed of a digital-to-analog converter (DAC), a single comparator, and some control logic and registers. When the analog voltage to be measured is present at the input to the comparator, the system control logic initially sets all bits to zero. Then the DAC's most significant bit (MSB) is set to 1, which forces the DAC output to 1/2 of full scale (in the case of a 10-V full-scale system, the DAC outputs 5.0 V). The comparator then compares the analog output of the DAC to the input signal, and if the DAC output is lower than the input signal, (the signal is greater than 1/2 full scale), the MSB remains set at 1. If the DAC output is higher than the input signal, the MSB resets to zero. Next, the second MSB with a weight of 1/4 of full scale turns on (sets to 1) and forces the

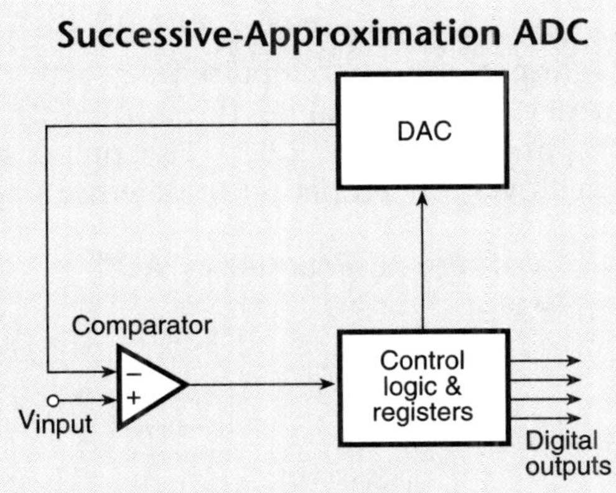

Fig. 2.01. Interestingly, this ADC uses a digital-to-analog converter and a comparator. The logic sets the DAC to zero and starts counting up, setting each following bit until it reaches the value of the measured input voltage. The conversion is then finished and the final number is stored in the register.

output of the DAC to either 3/4 full scale (if the MSB remained at 1) or 1/4 full scale (if the MSB reset to zero). The comparator once more compares the DAC output to the input signal and the second bit either remains on (sets to 1) if the DAC output is lower than the input signal, or resets to zero if the DAC output is higher than the input signal. The third MSB is then compared the same way and the process continues in order of descending bit weight until the LSB is compared. At the end of the process, the output register contains the digital code representing the analog input signal.

Successive approximation ADCs are relatively slow because the comparisons run serially, and the ADC must pause at each step to set the DAC and wait for its output to settle. However, conversion rates easily can reach over 1 MHz. Also, 12 and 16-bit successive-approximation ADCs are relatively inexpensive, which accounts for their wide use in many PC-based data acquisition systems.

Voltage-to-Frequency ADCs

Voltage-to-frequency ADCs convert the analog input voltage to a pulse train with the frequency proportional to the amplitude of the input. (See Figure 2.02.) The pulses are counted over a fixed period to determine the frequency, and the pulse counter output, in turn, represents the digital voltage.

Voltage-to-frequency converters inherently have a high noise rejection characteristic, because the input signal is effectively integrated over the counting interval. Voltage-to-frequency conversion is commonly used to convert slow and noisy signals. Voltage-to-frequency ADCs are also widely used for remote sensing in noisy environments. The input voltage is converted to a frequency at the remote location and the digital pulse train is transmitted over a pair of wires to the counter. This eliminates noise that can be introduced in the transmission lines of an analog signal over a relatively long distance.

Voltage-to-Frequency ADC

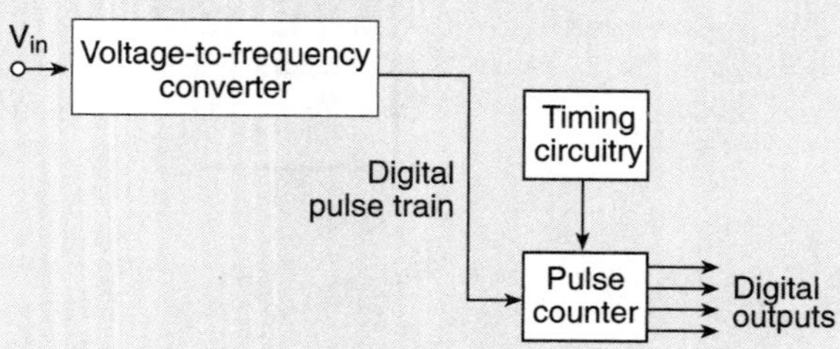

Fig. 2.02. Voltage-to-frequency converters reject noise well and frequently are used for measuring slow signals or those in noisy environments.

Integrating ADCs: Dual Slope

A number of ADCs use integrating techniques, which measure the time needed to charge or discharge a capacitor in order to determine the input voltage. A widely used technique, called dual-slope

Dual-Slope ADC Integration and Discharge Times

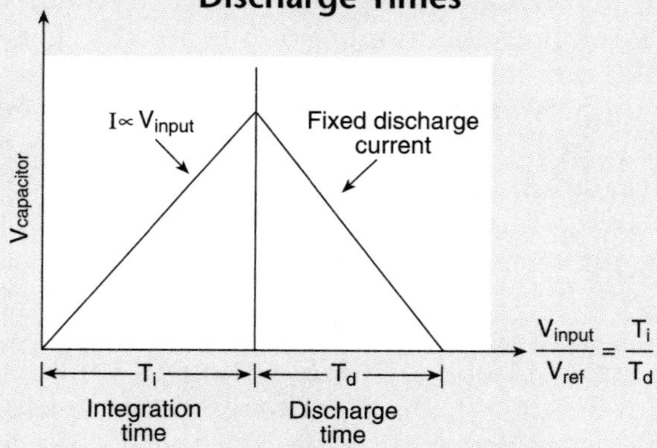

Fig. 2.03. Dual-slope integrating ADCs provide high-resolution measurements with excellent noise rejection. They integrate upward from an unknown voltage and then integrate downward with a known source voltage. They are more accurate than single slope ADCs because component errors are washed out during the de-integration period.

integration, is illustrated in Figure 2.03. It charges a capacitor over a fixed period with a current proportional to the input voltage. Then, the time required to discharge the same capacitor under a constant current determines the value of the input voltage. The technique is relatively accurate and stable because it depends on the ratio of rise time to fall time, not on the absolute value of the capacitor or other components whose values change over temperature and time.

Integrating the ADC input over an interval reduces the effect of noise pickup at the ac line frequency when the integration time is matched to a multiple of the ac period. For this reason, it is often used in precision digital multimeters and panel meters. Although 20-bit accuracy is common, it has a relatively slow conversion rate, such as 60 Hz maximum, and slower for ADCs that integrate over multiples of the line frequency.

Sigma-Delta ADCs

A sigma-delta ADC is another type of integrating ADC. It contains an integrator, a DAC, a comparator, and a summing junction. (See Figure 2.04.) Like the dual-slope ADC, it's often used in digital multimeters, panel meters, and data acquisition boards. Sigma-delta converters are relatively inexpensive primarily because they have a single-bit DAC, but they can obtain high-resolution measurements using oversampling techniques. Although the ADC works best with low-bandwidth signals (a few kHz), it typically has better noise rejection than many others, and users can set the integration time (albeit below 100 samples/sec).

Sigma-delta ADCs also require few external components. They can accept low-level signals without much input-signal conditioning circuitry for many applications, and they don't require trimming or calibration components because of the DAC's architecture The ADCs also contain a digital filter, which lets them

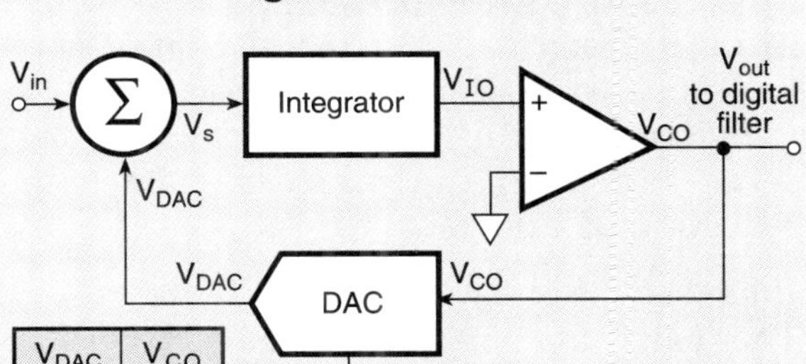

V_{DAC}	V_{CO}
$+V_{ref}$	1
$-V_{ref}$	0

Fig. 2.04. Integrating converters such as the sigma-delta ADC have both high resolution and exceptional noise rejection. They work particularly well for low-bandwidth measurements and reject high-frequency noise as well as 50/60 Hz interference.

work at a high oversampling rate without a separate anti-aliasing filter at the input. Sigma-delta ADCs come in 16 to 24-bit resolution, and they are economical for most data acquisition and instrument applications.

The principle of operation can be understood from the diagram. The input voltage V_{in} sums algebraically with the output voltage of the DAC, and the integrator adds the summing point output V_S to a value it stored previously. When the integrator output is equal to or greater than zero, the comparator output switches to logic one, and when the integrator output is less than zero, the comparator switches to logic zero. The DAC modulates the feedback loop, which continually adjusts the output of the comparator to equal the analog input and maintain the integrator output at zero. The DAC keeps the integrator's output near the reference voltage level. Through a series of iterations, the output signal becomes a one-bit data stream (at a high sample rate) that feeds a digital filter. The digital filter averages the series of logic ones and zeros, determines the bandwidth and settling time, and outputs multiple-bit data. The digital low-pass filter then feeds the decimation filter, which in turn, decreases the sample rate of the multi-bit data stream by a factor of two for each stage within the filter. For example, a seven-stage filter can provide a sample-rate reduction of 128.

Improved accuracy

The digital filter shown in Figure 2.05 inherently improves the ADC's accuracy for ac signals in two ways. First, when the input signal varies (sine wave input) and the system samples the signal at several times the Nyquist value (refer to page 17), the integrator becomes a low-pass filter for the input signal, and a high-pass filter for the quantization noise. The digital filter's averaging function then lowers the noise floor even further, and combined with the decimation filter, the data stream frequency at the output is reduced. For example, the modulator loop frequency could be in the MHz region, but the output data would be in the kHz region. Second, the digital filter can be notched at 60 Hz to eliminate power line frequency interference.

Sigma-Delta ADC With Digital Filter and Decimator Stage

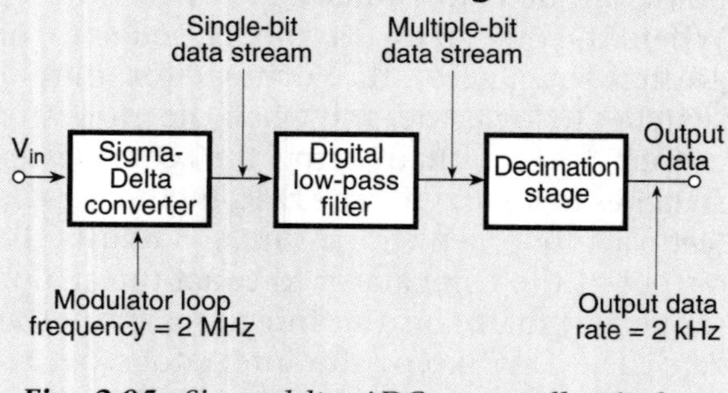

Fig. 2.05. Sigma-delta ADCs are well suited to high-resolution acquisition because they use oversampling and often combine an analog modulator, a digital filter, and a decimator stage. The low-pass digital filter converts the analog modulator output to a digital signal for processing by the decimator.

The output data rate from the decimation filter is lower than the initial sample rate but still meets the Nyquist requirement by saving certain samples and eliminating others. As long as the output data rate is at least two times the bandwidth of the signal, the decimation factor or ratio M can be any integer value. For example, if the input is sampled at f_S, the output data rate can be f_S/M without losing information. This technique provides more stable readings. (Refer to the table in Figure 2.06 for ADC comparisons.)

Table of ADC Attributes

ADC Type	Typical Resolution	Typical Conversion Rate*/Frequency
Sigma-Delta	16–24 bit	1 sps–128 ksps
Successive Approximation	8–16 bit	10 ksps–2 Msps
Voltage-to-Frequency	8–12 bit	1 Hz–4 MHz**
Integrating	12–24 bit	1 sps–1 ksps**

* sps = samples per second
** With line cycle rejection

Fig. 2.06. Table of ADC attributes

ACCURACY AND RESOLUTION

Accuracy is one of the most critical factors to consider when specifying an ADC for test and measurement applications. Unfortunately, it's often confused with resolution, and although related, they are distinctly different. Both topics are discussed in this section in some detail, as well as their relationship to calibration, linearity, missing codes, and noise.

Accuracy versus Resolution

Every ADC measurement contains a variety of unavoidable, independent errors that influence its accuracy. When σ_i represents each independent error, the total error can be shown as:

$$\sigma_{total} = \sqrt{\Sigma_i \sigma_i^2}$$

This equation includes a variety of errors such as sensor anomalies, noise, amplifier gain and offset, ADC quantization (resolution error), and other factors.

Quantization Error

In a theoretically perfect ADC, any particular analog voltage measured should be represented by a unique digital code, accurate to an infinite number of digits. (See Figure 2.07A.) But in a real ADC, small but finite gaps exist between one digital number and a consecutive digital number, and the amount depends on the smallest quantum value that the ADC can resolve. In the case of the 12-bit converter covering a 10-Vdc range, for example, that quantum value is 2.44 mV, the LSB. In other words, the input analog voltage range is partitioned into a discrete number of values that the converter can measure, which is also the ADC's resolution. The quantization error in this case is specified to be no more than half of the least significant bit (LSB). For the 12-bit ADC, the error is ±1.22 mV (0.0122%). Such ADC errors are typically specified in three ways: the error in LSBs, the voltage error for a specified range, and the % of reading error. Most ADCs are not as accurate as their specified resolution, however, because other errors contribute to the overall error such as gain, linearity, missing codes, and offset. (See Figures 2.07B, C, D, and E, respectively). Nonetheless, the accuracy of a good ADC should approach its specified resolution. When an ADC manufacturer provides calibra-

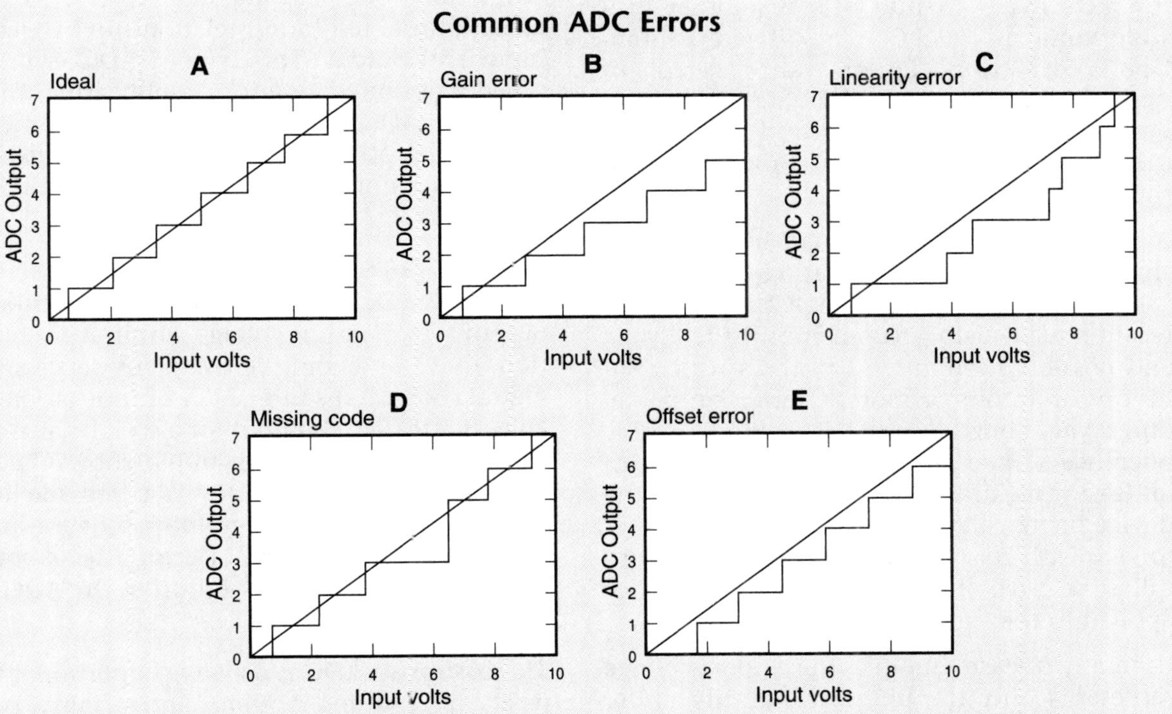

Fig. 2.07. The straight line in each graph represents the analog input voltage and the perfect output voltage reading from an ADC with infinite resolution. The step function in Graph A shows the ideal response for a 3-bit ADC. Graphs B, C, D, and E show the effect on ADC output from the various identified errors.

tion procedures, offset and gain errors usually can be reduced to negligible levels, however, linearity and missing-code errors are more difficult or impossible to reduce.

ADC Accuracy versus System Accuracy

Calibration

ADCs may be calibrated with hardware, software, or a combination of the two. Calibration in this case means adjusting the gain and offset of an ADC channel to obtain the specified input-to-output transfer function. In a hardware configuration, for example, the instrumentation amplifier driving the ADC has its offset and gain adjusted with trim pots, and changing the ADC's reference voltage changes its gain. In hardware/software calibrations, the software instructs DACs to null offsets and set full-scale voltages. Lastly, in a software calibration, correction factors are stored in nonvolatile memory in the data acquisition system or in the computer and are used to calculate the correct digital value based on the readings from the ADC.

ADCs are factory calibrated before being shipped, but time and operating temperature can change the settings. ADCs need to be recalibrated usually after six months to a year, and possibly more often for ADCs with resolutions of 16 bits or more. Calibration procedures vary, but all usually require a stable reference source and an indicating meter of (at least three times) greater accuracy than the device being calibrated. Offset is typically set to zero with zero input, and the gain is set to full scale with the precise, full-scale voltage applied to the input.

In many measurements, the input voltage represents only the physical quantity under test. Consequently, system accuracy might be improved if the complete measurement system is calibrated rather than its individual parts. For example, consider a load cell with output specified under a given load and excitation voltage. Calibrating individual parts means that the ADC, load cell, and excitation source accuracy tolerances are all added together. With this approach, the error sources of each part are added together and generate a total error that is greater than the error that can be achieved simply by calibrating the system with a known precision load and obtaining a direct relationship between the input load and ADC output.

Linearity

When the input voltage and the ADC output readings deviate from the diagonal line (representing infinite resolution) more than the ideal step function shown in Figure 2.07A, the ADC error is nearly impossible to eliminate by calibration. The diagonal line represents an ideal, infinite-resolution relationship between input and output. This type of ADC error is called a nonlinearity error. Nonlinearities in a calibrated ADC produce the largest errors near the middle of the input range. As a rule of thumb, nonlinearity in a good ADC should be one LSB or less.

Missing Codes

A quality ADC should generate an accurate output for any input voltage within its resolution, that is, it should not skip any successive digital codes. But some ADCs cannot produce an accurate digital output for a specific analog input. Figure 2.07D, for example, shows that a particular 3-bit ADC does not provide an output representing the number four for any input voltage. This type of error affects both the accuracy and the resolution of the ADC.

Noise

The cost of an ADC is usually proportional to its accuracy, number of bits, and stability. But even the most expensive ADC can compromise accuracy when excessive electrical noise interferes with the measured signal, whether that signal is in millivolts or much larger.

For example, many ADCs that reside on cards and plug into a PC expansion bus can encounter excessive electrical noise that seriously affects their accuracy, repeatability, and stability. But an ADC does not have to be connected directly to the bus within the computer. An ADC mounted in an external enclosure often solves the problem. It can communicate with the computer over an IEEE 488 bus, Ethernet, serial port, or parallel port.

When there is no choice but to locate an ADC inside the computer, however, check its noise level. Connecting the ADC's input terminal to the signal common terminal should produce an output of zero volts. If it still reads a value when shorted, the noise is being generated on the circuit card and will interfere with the desired input signal. More critical diagnostics are necessary when using an external power supply because noise also can arise from both the power supply and the input leads.

Noise Reduction and Measurement Accuracy

One technique for reducing noise and ensuring measurement accuracy is to eliminate ground loops, that is, current flowing in the ground connection between different devices. Ground loops often occur when two or more devices in a system, such as a measurement instrument and a transducer, are connected to ground terminals at different physical locations. Slight differences in the actual potential of each ground point generate a current flow from one device to the other. This current, which often flows through the low potential lead of a pair of measurement wires generates a voltage drop that appears as noise and measurement inaccuracy at the signal conditioner or ADC input. When at least one device can be isolated, such as the transducer, then the offending ground path is open, no current flows, and the noise or inaccuracy is eliminated. Optical isolators, special transformers, and differential input operational amplifiers at the signal conditioner or ADC input can provide this isolation.

ADC Noise Histograms

ADC manufacturers frequently verify their device's accuracy (effect of non-linearities) by running a code density test. They apply a highly accurate sine wave signal (precision amplitude and frequency) to the device and using a histogram for analysis, generate a distribution of digital codes at the output of the converter. A perfect ADC would produce only one vertical bar in the histogram for the specified input frequency and amplitude because it measured only one value for every sample. But because of the ADC's inherent non-linearities, it produces a distribution of bars on either side representing digital words sorted into different code bins. Each bin is labeled for a single digital output code and it contains the count of its occurrence, or the number of times that code showed up in the output. (See Figure 2.08.)

When n represents the ADC's bit resolution, 2^n bins are required. The width of each code bin should be $FSR/2^n$ where FSR is the full-scale range of the ADC. The probability density function may be determined from this data. A large number of samples must be taken, depending on the ADC's bit size, for the histogram test to be meaningful. The more bits the ADC contains, the higher the number of samples required, which could be as much as 500,000 samples.

ENOB: EFFECTIVE NUMBER OF BITS

Although an ADC's accuracy is crucial to a data acquisition system's accuracy, it's not the final word. A widely used and practical way of determining overall measurement accuracy is accomplished with what's called an Effective Number of Bits (ENOB) test. The ENOB may very well demonstrate that the actual system

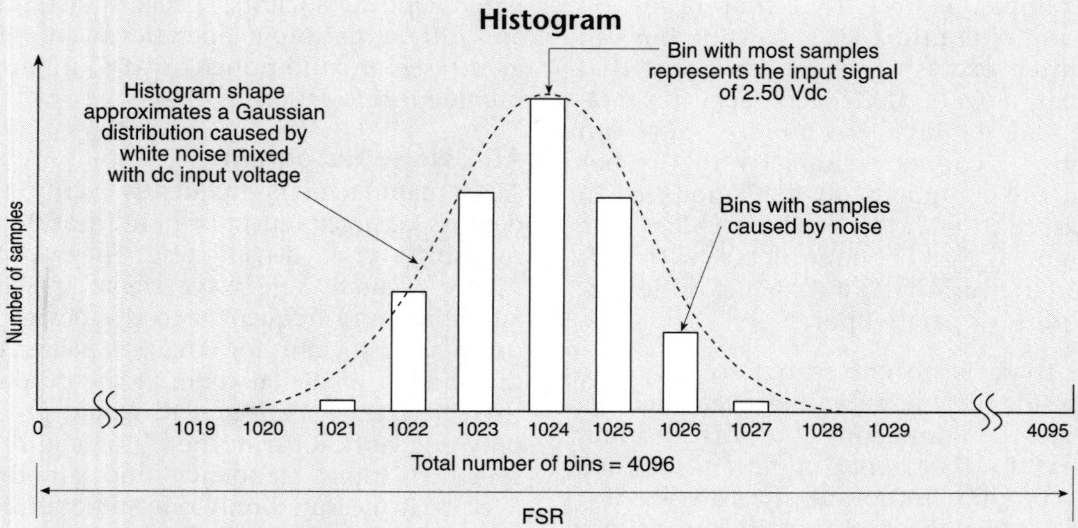

Fig. 2.08. *The histogram illustrates how 12-bit ADC samples in a set were distributed among the various codes for a 2.5-V measurement in a FSR (full-scale range) of 10-V. Most codes intended for the 1024 bin representing 2.5 V actually ended up there, but others fell under a gaussian distribution due to white noise content.*

measurement accuracy is something less than the ADC bit accuracy specifications. For example, an ADC may be specified as a 16-bit device, but the results of a specific type of standard test may show that its performance is actually closer to an ideal 13-bit system. However, 13 bits may be more than adequate for the application.

The ENOB test takes into account all the circuitry from input terminals to the data output, which includes the effects of the ADC, multiplexer, and other analog and digital circuits on the measurement accuracy. It also includes the signal-to-noise ratio, SNR, or the effect of any noise signals induced into the system from any source.

The ENOB Test

The ENOB test evaluates the data acquisition system as it performs in a real-world application when used with the manufacturers recommended cables, connectors, and connections. It considers the front end of the data acquisition system: the ADC, multiplexer, programmable gain amplifier, and sample-and-hold amplifiers. All of these circuits affect the digitized output. Any non-linearities, noise, distortion, and other anomalies that sneak into the front end can reduce the system's accuracy, and it's not enough to test only one channel in a multi-channel system. Some errors creep up from the effect of one channel on another through cross talk.

To measure the ENOB, set up a precision sine-wave signal generator and connect it's output to the input of one analog input channel. The generator itself should produce little noise and distortion. Set the signal generator output amplitude to just under the maximum specified input range of the board. Set the generator to the maximum frequency that the system is specified to measure. Next, ground the input terminals of the adjacent channel. Run the system at its maximum rated speed. Sample the test signal and then the grounded input. Capture 1024 samples on each input and run the samples through an FFT algorithm to compute the ENOB.

The test measures the effects of slewing, harmonic distortion, analog circuits, ADC accuracy, noise pickup, channel cross talk, integral and differential non-linearity, and offset between channels. Although ENOB provides a good benchmark of the systems capability and accuracy, it is not a specification; it is not a replacement for SNR and other error specifications provided by the manufacturer. However, systems may be compared with ENOB tests when all are measured under the same set of conditions.

ADC OUTPUT AVERAGING BENEFITS

Improved Accuracy

A paradox that arises from averaging the output of an integrating-type ADC is that the measurement system theoretically can obtain higher accuracy for a signal embedded in noise than a signal free of noise. How this can be true comes from the way a signal is mathematically averaged. For example, with a single dc signal, averaging the output always provides the same result with no apparent change in accuracy (not considering the affects of calibration). But for a varying input signal, such as a sine wave, a large number of samples yield a Gaussian distribution which can be accurately defined with a more precisely established peak for the wave. But the samples must not all cluster around a specific portion of the sampled wave. To get a true distribution, the ADC must sample at a slower rate than the fluctuations and be out of synch with them. This technique finds a general average, not a local average. Thus, in this way signal averaging increases the system's measurement resolution.

More Stable Readings

Some systems actually introduce a random noise signal called dither into an otherwise clean ADC input to take advantage of the averaging function for increasing accuracy and signal stability. The technique also enables an ADC with a smaller number of bits to obtain the resolution of an ADC with more bits without losing accuracy. Each signal sweep must capture a different random value at each point in time. Then the theoretical ADC average from this distribution will remain at zero over a sufficiently large sample window. For example, if 16 values are averaged, then it has 16 times more possible values than the direct non-averaged output. This technique increases the effective ADC resolution by 4 bits. The noise makes it work.

ADC signal averaging is so important a technique that it is used on digital recordings. In the early days of digital audio recording development, systems lacked ADC output averaging. As a result, a musical note would decay into an annoying buzz because not all bits in the ADC were enabled. The output waveforms were distorted and the ear could not filter it out, but ADC signal averaging totally eliminated the problem.

Chapter 3

MULTIPLEXING AND SAMPLING THEORY

THE ECONOMY OF MULTIPLEXING

Sampled-Data Systems

An ideal data acquisition system uses a single ADC for each measurement channel. In this way, all data are captured in parallel and events in each channel can be compared in real time. But using a multiplexer, Figure 3.01, that switches among the inputs of multiple channels and drives a single ADC can substantially reduce the cost of a system. This approach is used in so-called sampled-data systems. The higher the sample rate, the closer the system mimics the ideal data acquisition system. But only a few specialized data acquisition systems require sample rates of extraordinary speed. Most applications can cope with the more modest sample rates typically offered by mainstream data acquisition systems.

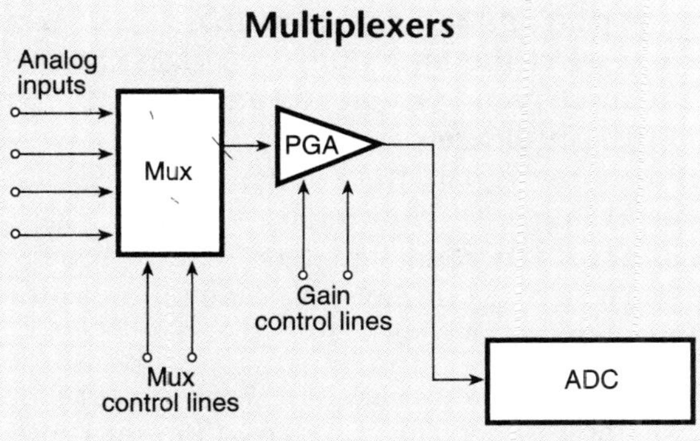

Fig. 3.01. The multiplexer (MUX) is a fast switch that sequentially scans numerous input-signal channels and directs them in a preprogrammed manner to a single ADC for digitizing. This approach saves the cost of using an ADC for each channel.

Solid State Switches vs. Relays

A multiplexer is an array of solid-state switches or electromechanical relays connected to several input channels. Although both approaches are used in a wide variety of applications, neither one is perfect; each type comes with various advantages and disadvantages. Electromechanical relays, for example, are relatively slow, about 1,000 samples/sec or less for the fastest reed relays, but they can handle large input voltages, and some can isolate voltages of several kV. A relay's size and contact type determine its current carrying capacity. For instance, laboratory instrument relays typically switch up to 3 A, while industrial applications use larger relays to switch higher currents, often 5 to 10 A.

Solid-state switches, on the other hand, are much faster than relays and can reach sampling rates of several MHz. However, these devices can't handle inputs higher than 25 V, and they are not well suited for isolated applications. Moreover, solid-state devices are typically limited to handling currents of only one mA or less.

Another characteristic that varies between mechanical relays and solid-state switches is called ON resistance. An ideal mechanical switch or relay contact pair has zero ON resistance. But real devices such as common reed-relay contacts are 0.010 Ω or less, a quality analog switch can be 10 to 100 Ω, and an analog multiplexer can be 100 to 2,500 Ω per channel.

The ON resistance adds directly to the signal source impedance and can affect the system's measurement accuracy if not compensated.

Analog switching devices have another undesirable characteristic called charge injection. This means that a small portion of input-gate drive voltage is coupled to the analog input signal and manifests as a spike in the output signal. This glitch produces measurement errors and can be seen riding on the input signal when the source impedance is too high. A compensating circuit can minimize the effects of charge injection, but the most effective method is to keep source impedance as low as possible to prevent it from developing in the first place.

Channel-to-channel cross talk is another nonideal characteristic of analog switching networks, especially integrated circuit multiplexers. Cross talk develops when the voltage applied to any one channel affects the accuracy of the reading in another channel. Conditions are optimum for cross talk when signals of relatively high frequency and high magnitude such as 4 to 5 V signals are connected to one channel while 100 mV signals are connected to an adjacent channel. High frequency multiplexing also exacerbates cross talk because the signals couple through a small capacitance between switch channels. Low source impedance minimizes the cross talk and eliminates the charge injection.

Speed

Multiplexing reduces the rate at which data can be acquired from an individual channel because of the time-sharing strategy between channels. For example, an ADC that can sample a single channel at 100 kHz is limited to a 12.5-kHz/channel sampling rate when measuring eight channels.

Unfortunately, multiplexing can introduce yet other problems. For instance, the multiplexer's high source impedance can combine with stray capacitance to increase settling time and generate cross talk between channels. Multiplexer impedance itself also can degrade signals. A solid-state multiplexer with an impedance of tens or hundreds of ohms is worse than a relay with a typical resistance of 0.010 Ω or less.

In spite of these negative issues, the advantages of multiplexing outweigh its disadvantages, and it has become a widely used technique to minimize cost without compromising performance. Because the measurement errors are known and specified, they can be compensated at each stage of the data acquisition system to ensure high accuracy at the output.

Sequence vs. Software-Selectable Ranges

Most data acquisition systems accommodate a variety of input ranges, although the manner in which they do so varies considerably. Some data acquisition systems allow the input range to be switched or jumper selected on the circuit board. Others provide software-selectable gain. This is more convenient, but a distinction should be made between data acquisition systems whose channels must all have the same gain and other systems that can sequentially select the input range for each channel. The more useful system accepts different input ranges on different channels, especially when measuring signals from different transducers. For example, thermocouples and strain gages require input ranges in tens of mV and use special signal conditioners, while other sensors might output several volts.

A data acquisition system with a software-selectable range can measure different ranges on different channels (but at a relatively slow rate) with a command to change the gain between samples. But the technique has two problems. First, it is relatively slow. That is, issuing a software command to change the gain of a programmable-gain amplifier (PGA) can take tens or hundreds of ms, lowering the system's sample rate to several Hz. Second, the speed of this sequence is often indeterminate due to variations in PC instruction cycle times. Cycling through the sequence continuously generates samples with an uneven (and unknown) spacing in time. This complicates time-series analysis and makes FFT analysis impossible because its algorithm requires evenly spaced samples.

A better implementation hosts a sequencer that sets the maximum acquisition rate and controls both channel selection and associated amplifier gain at random. For example, one widely used data acquisition system running at 100 kHz and 1 MHz uses software selectable channel gain and sequencing. (See Figure 3.02.) The 100-kHz system provides a 512-location scan sequencer that lets operators use software to select each channel and its input amplifier gain for both the built-in and expansion channels. Each scan group can be repeated immediately or at programmable intervals. The sequencer circuitry overcomes a drastic reduction in the scan rate for expansion channels, a major limitation encountered with many plug-in data acquisition boards.

All channels are scanned, including expansion channels, at 100 kHz, (10 µs per channel). (See Figure 3.03.) Digital inputs also can be scanned using the same scan sequence employed for analog inputs, enabling the time correlation of acquired digital data to acquired analog data. Such systems permit each scan group (containing up to 512 channel/gain combinations) to be repeated immediately or programmed to intervals of up to 12 hours. Within each scan group, consecutive channels are measured at a fixed 10-µs per channel rate.

FUNDAMENTAL CONCEPTS

Sampling Rates

When ADCs convert an analog voltage to a digital representation, they sample the value of the measured variable several times per second. Steady or slowly changing dc voltages might require sampling rates of only

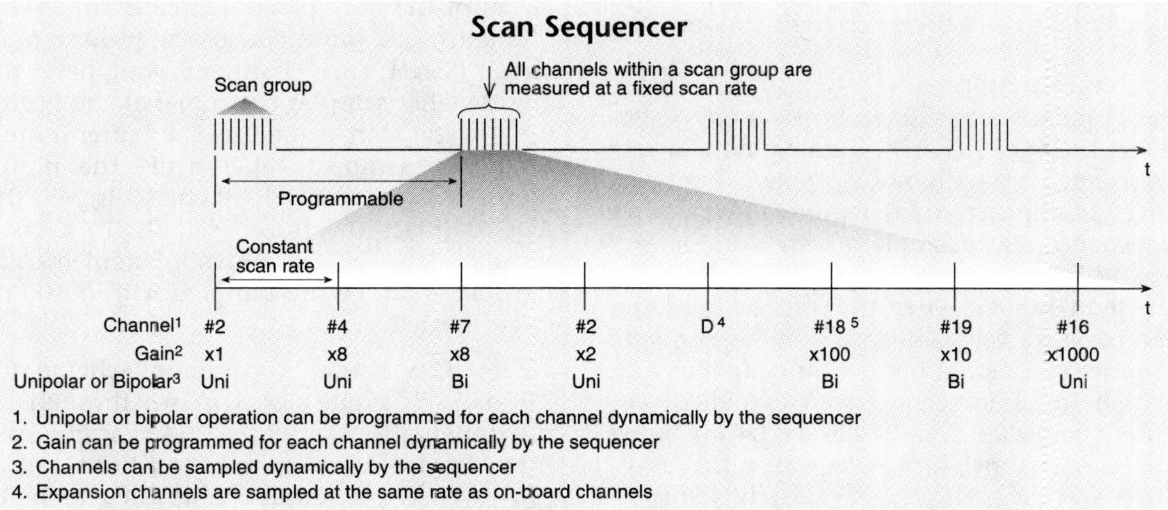

Fig. 3.02. This is an example of a 512-location scan sequencer operating in a 100-kHz data acquisition system. The sequencer can program each channel dynamically for unipolar or bipolar operation. The sequencer also can program the gain for each channel on the fly and change the order in which the channels are scanned.

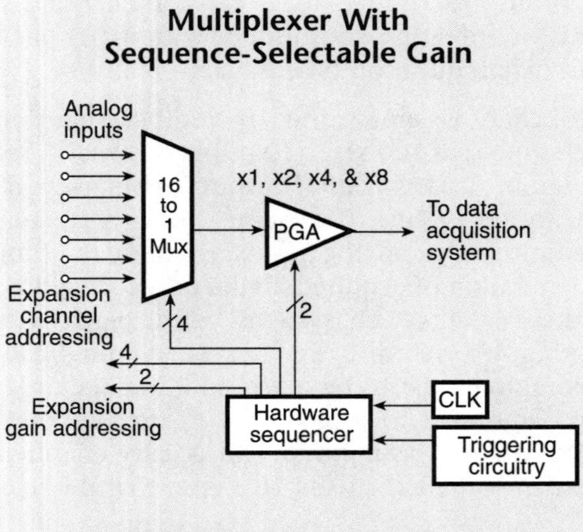

Fig. 3.03. A multiplexer with sequence-selectable gain lets the data system operate at its maximum acquisition rate.

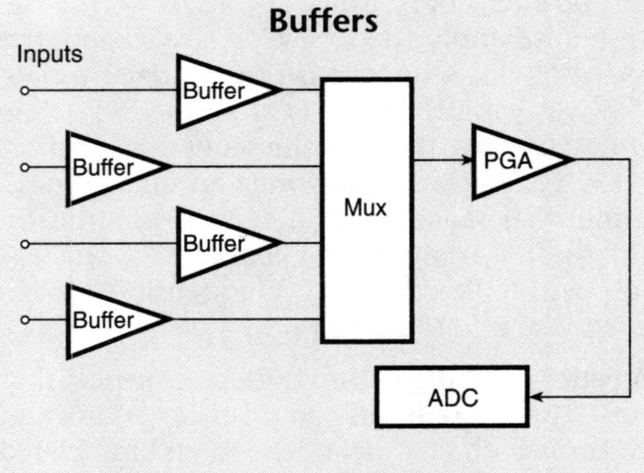

Fig. 3.04. Buffering signals ahead of the multiplexer increases accuracy, especially with high-impedance sources or fast multiplexers

several Hz, but measuring various ac and sine waves is different. Enough samples/sec must be made to ensure that the wave being measured in the continuous-time domain can be faithfully reproduced in both the continuous-time domain and the discrete-time domain.

Source Impedance

Most signal sources have impedances of less than 1.5 kΩ, so such a maximum source impedance is usually not a problem. However, faster multiplexer rates require lower source impedances. For example, a 1-MHz multiplexer in a 12-bit system requires a source impedance less than 1.0 kΩ. When the source impedance exceeds this value, buffering is necessary to improve accuracy. A buffer is an amplifier with a high input impedance and extremely low output impedance. (See Figure 3.04.) A buffer on each channel located between the transducer and the multiplexer ensures higher accuracies by preventing the multiplexer's stray capacitance from discharging through the impedance of the transducer.

Sample and Hold ADCs

Time Skew

A multiplexed ADC measurement introduces a time skew among channels, because each channel is sampled at a different time. Some applications can't tolerate this effect. But a sample and hold circuit placed on each input ahead of the multiplexer remedies time-skew problems. In a simultaneous sample and hold circuit (SS&H) each channel is equipped with a buffer that samples the signal at the beginning of the scan sequence. The buffer output holds the sampled value while the multiplexer switches through all channels, and the ADC digitizes the frozen signals. In a good simultaneous sample and hold implementation, all channels are sampled within 100 ns of each other.

Figure 3.05 shows a common scheme for SS&H. Each input signal passes through an instrumentation amplifier (IA), a low-pass filter, and into a sample-and-hold buffer (S/H). When the sample enable line goes high, each S/H samples its input signal and holds it while the multiplexer switches through the readings. This scheme ensures that all the

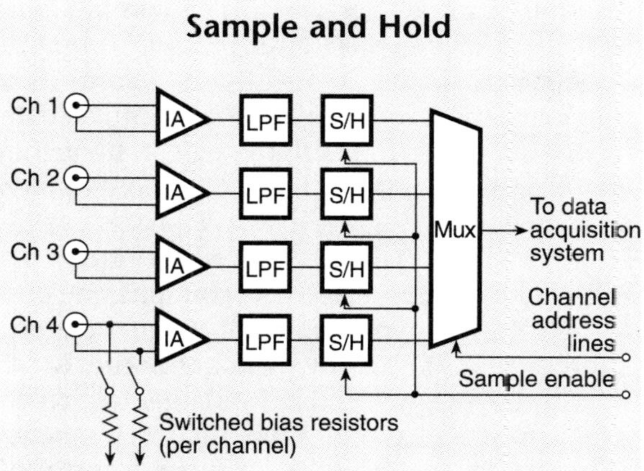

Fig. 3.05. A four-channel multiplexer with low-pass filters and simultaneous sample-and-hold circuits ensures that all the samples are taken within a few ns of each other.

samples are taken within 50 ns of each other, even with up to 256 simultaneous channels connected to a single instrument.

Nyquist Theorem

Transforming a signal from the time domain to the frequency domain requires the application of the Nyquist theorem. The Nyquist sampling theorem states that if a signal only contains frequencies less than cutoff frequency, f_C, all the information in the signal can be captured by sampling it at a minimum frequency of $2f_C$. This means that capturing a signal with a maximum frequency component of f_{max} requires that it must be sampled at $2f_{max}$ or higher. However, common practice dictates that while working in the frequency domain, the sampling rate must be set more than twice and preferably between five and ten times the signal's highest frequency component. Waveforms viewed in the time domain are usually sampled 10 times the frequency being measured to faithfully reproduce the original signal and retain accuracy of the signal's highest frequency components.

Aliasing and Fourier Transforms

When input signals are sampled at less than the Nyquist rate, ambiguous signals that are much lower in frequency than the signal being sampled can appear in the time domain. This phenomenon is called aliasing. For example, Figure 3.06 shows a 1-kHz sine wave sampled at 800 Hz. The reconstructed or transformed frequency of the sampled wave is much too low and not a true representation of the original. If the 1-kHz signal were sampled at 1,333 Hz, a 333-Hz alias signal would appear. Figure 3.07, on the other hand, shows the signals when sampling the same 1-kHz wave at more than twice the input frequency or 5 kHz. The sampled wave now appears closer to the correct frequency.

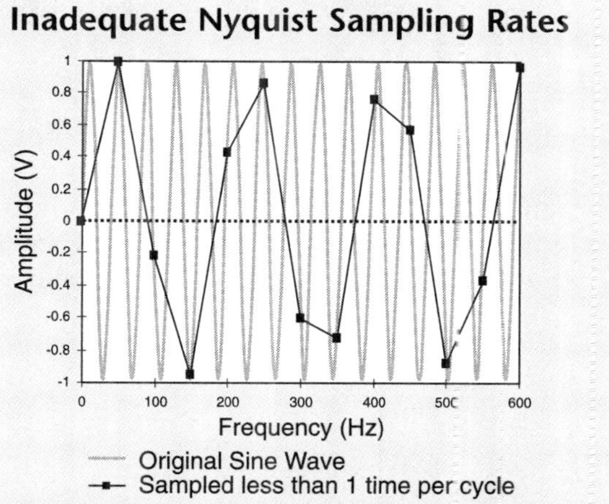

Fig. 3.06. When sampling an ac signal at less than two times the Nyquist frequency, the original waveform cannot be reproduced faithfully.

Conversely, input frequencies of half or more of the sampling rate will also generate aliases. To prevent these aliases, a low-pass, anti-aliasing filter is used to remove all components of these input signals. The filter is usually an analog circuit placed between the signal input terminals and the ADC. Although the filter eliminates the aliases, it also prevents any other signals from passing

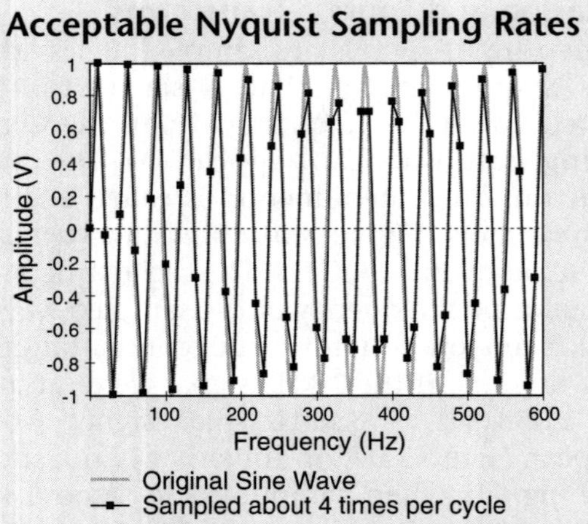

Fig. 3.07. When ac inputs are sampled more than twice the Nyquist frequency of the sine wave, the frequency content of the signal is preserved, and all the Fourier components of the periodic waveform are recovered.

through that are above the stop band of the filter, whether they were wanted or not. In other words, when selecting a data acquisition system, make certain that the per channel sampling frequency is more that twice the highest frequency intended to be measured.

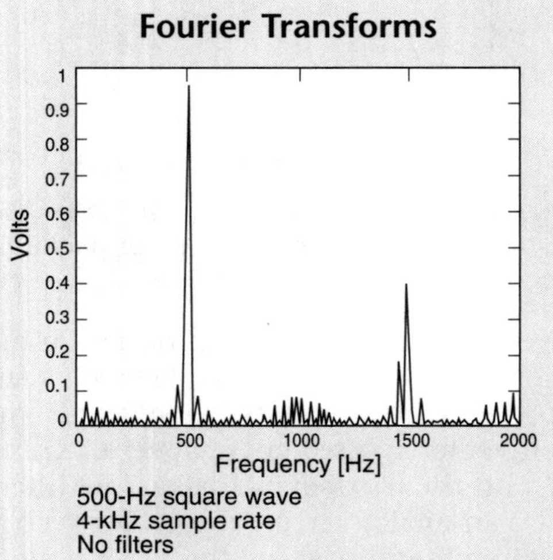

Fig. 3.08. The Fourier transform of a 500-Hz square wave is sampled at 4 kHz with no filtering.

Another example of aliasing is shown in Figure 3.08 for a square wave after passing through a Fourier transform. A Fourier transform is a frequency spectrum display of the sampled data. It shows how much energy at a given frequency is in a particular signal. For the purpose of illustration, assume that the example processes only frequencies under 2 kHz. Ideally, a Fourier transform of a 500-Hz square wave contains one peak at 500 Hz, the fundamental frequency, and another at 1,500 Hz, the third harmonic, which is 1/3 the height of the fundamental. Figure 3.08, however, shows how higher frequency peaks are aliased into the Fourier transform's low-frequency range. The low-pass filter with cutoff at 2 kHz shown in Figure 3.09 removes most of the aliased peaks.

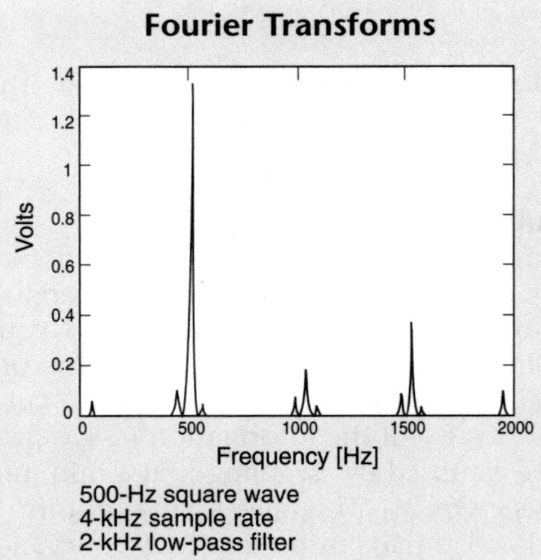

Fig. 3.09. The Fourier Transform of a 500-Hz square wave is sampled at 4 kHz with a low-pass filter cutoff at 2 kHz.

When the sampling rate increases to four times the highest frequency wanted, the Fourier transform in the range of interest looks even better. Although a small peak remains at 1 kHz, it is most likely the result of an imperfect square wave rather than an effect of aliasing. (See Figure 3.10).

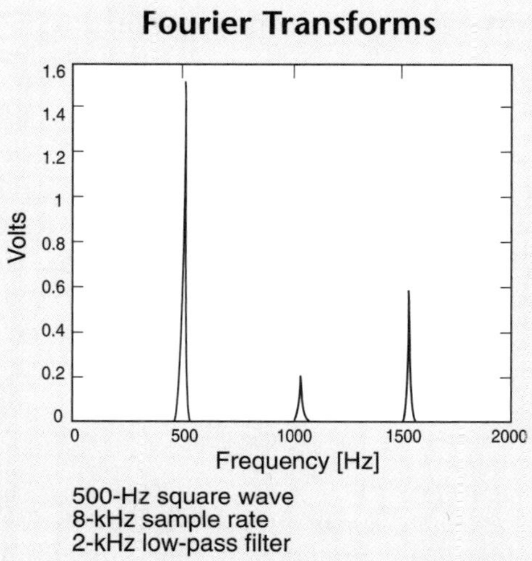

Fig. 3.10. The Fourier Transform of a 500-Hz square wave is sampled at 8 kHz with a low-pass filter cutoff at 2 kHz.

Discrete Fourier Transform

When ac signals pass through a time-invariant, linear system, their amplitude and phase components can change but their frequencies remain intact. This is the process that occurs when the continuous time domain ac signal passes through the ADC to the discrete time domain. Sometimes, more useful information can be obtained from the sampled data by analyzing them in the discrete time domain with a Fourier Series rather than reconstructing the original signal in the time domain.

The sampled data pass through a Fourier transform function to cull out the fundamental and harmonic frequency information. The amplitude of the signal is displayed in the vertical axis, and the frequencies measured are plotted on the horizontal axis.

Windowing

Real-time measurements are taken over finite time intervals. In contrast, Fourier transforms are defined over infinite time intervals, so limiting the transform to a discrete time interval produces sampled data that are only approximations. Consequently, the resolution of the Fourier transform is limited to approximately 1/T where T is the finite time interval over which the measurement was made. Fourier transform resolution can only be improved by sampling for a longer interval.

A finite time interval used for a Fourier Transform also generates spurious oscillations in the transform's display as shown in Figure 3.11. From a mathematical viewpoint, the signal that's instantaneously turned on at the beginning of the measurement and then suddenly turned off at the end of the measurement produces the spurious oscillations. These spurious oscillations are usually eliminated with a function called windowing, that is, multiplying the sampled data by a weighting function. A window function that rises gradually from zero decreases the spurious oscillations at the expense of a slight loss in triggering resolution. Many possible window functions may be used, including Hanning, Hamming, Blackman, Rectangular, and Bartlett. (See Figures 3.12, 3.13, and 3.14.)

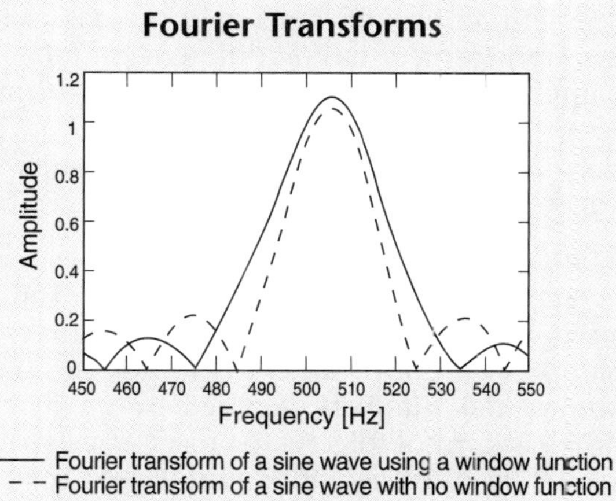

Fig. 3.11. A Fourier Transform is shown with and without a window function.

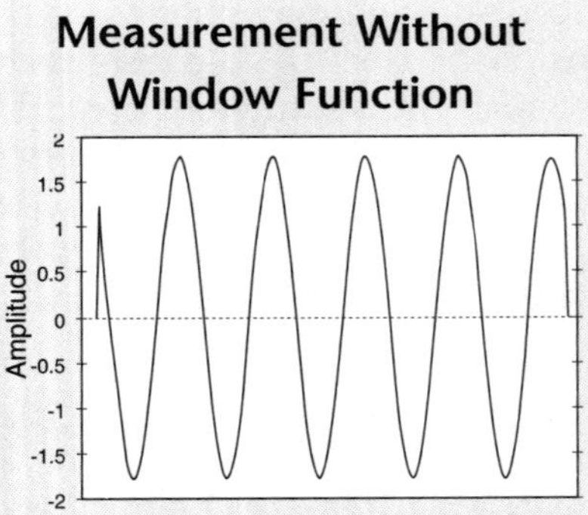

Fig. 3.12. The frequency information obtained from a measurement can be incorrect when the waveform contains abrupt beginning or ending points.

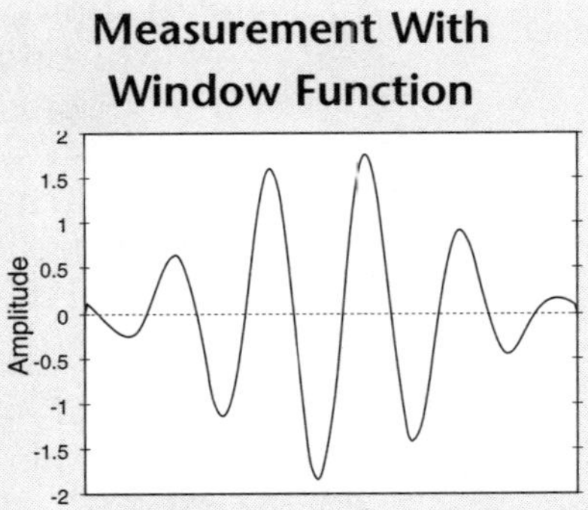

Fig. 3.13. In this case, the frequency information is preserved when multiplied by a windowing function. It minimizes the effect of irregularities at the beginning and end of the sample segment.

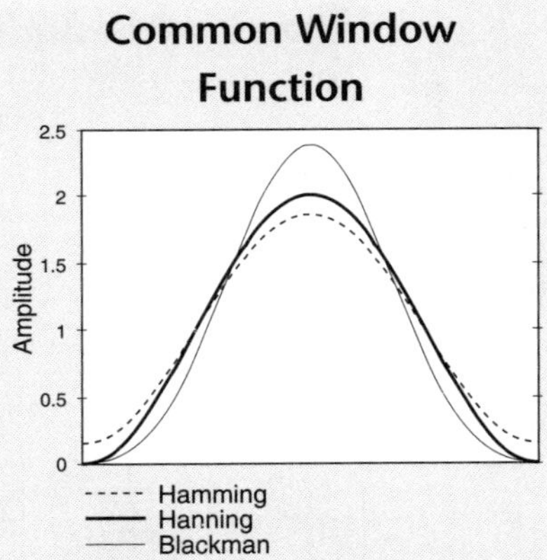

Fig. 3.14. The three most common types of window functions include Hamming, Hanning, and Blackman.

Fast Fourier Transforms

The Fast Fourier Transform (FFT) is so common today that FFT has become an imprecise synonym for Fourier transforms in general. The FFT is a digital algorithm for computing Fourier transforms of data discretely sampled at a constant interval. The FFT's simplest implementation requires 2^n samples. Other implementations accept other special numbers of samples. If the data set to be transformed has a different number of samples than required by the FFT algorithm, it is often padded with zeros to meet the required number. Sometimes the results are inaccurate, but most often they are tolerable.

Standard Fourier Transforms

A Standard Fourier Transform (SFT) can be used in applications where the number of samples cannot be arranged to fall on one of the special numbers required by an FFT, or where it cannot tolerate the inaccuracies introduced by padding with zeros. The SFT is also suitable where the data are not sampled at evenly spaced intervals or where sample points are missing. Finally, the SFT can be used to provide more closely spaced points in the frequency domain than can be obtained with an FFT. In an FFT, adjacent points are separated by 1/T, the inverse of the time interval over which the measurement was made.

Many standard numerical integration techniques exist for computing SFTs from sampled data. Any other technique selected for the problem at hand probably will be much slower than an FFT of a similar number of points. This is becoming less of an issue, however, as the speed of modern computers increases.

Digital vs. Analog Filtering

Digital filtering requires three steps. First, the digital signal is subjected to a Fourier transform. Then the signal amplitude in the frequency domain is multiplied by the desired frequency response. Finally the transform signal is inverse Fourier transformed back into the time domain. Figure 3.15 shows the effect of digital filtering on a noisy signal. The solid line represents the unfiltered signal while the two dashed lines show the affect of different digital filters. Digital filters have the advantage of being easily tailored to any frequency response without introducing phase error. However, one disadvantage is that a digital filter cannot be used for anti-aliasing.

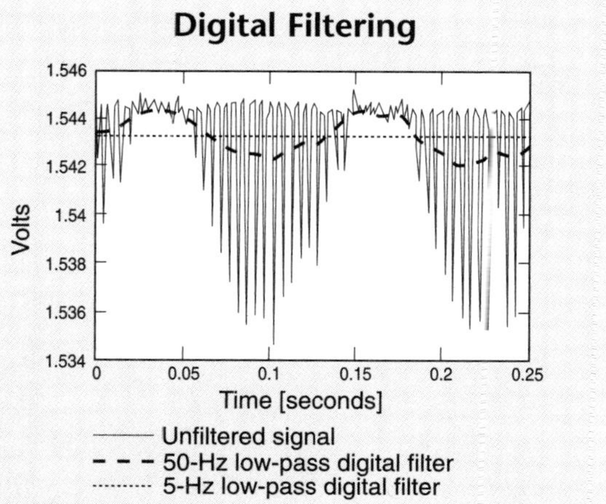

Fig. 3.15. The solid line is the noisy, unfiltered signal, and the two dashed lines show the same signal at the output of the 5 and 50-Hz low-pass digital filters.

In contrast to digital filters, analog filters can be used for anti-aliasing. But changing the frequency response curves is more difficult because all analog filters introduce some phase error.

Settling Time

Source impedance and stray capacitance influence the multiplexer input settling time. Their effect can be predicted with a simple equation:

EQN. 3.01 Time Constant

$T = RC,$
Where:
T = time constant, s
R = source impedance, Ω
C = stray capacitance, F

For example, determine the maximum tolerable source impedance for a 100-kHz multiplexer. The time between measurements on adjacent channels in the scan sequence is 10 µs. During time T = RC, the voltage error decays by a factor of 2.718, or one time constant, T. But a time interval 10 times longer (10 T) is required to reduce the error to 0.005%. Consequently, a fixed time of 10 µs between scans (T_{scan}) and an error of 0.005% requires that T = 1 µs. But in a typical multiplexed data acquisition system, this is insufficient settling time, so the data would continue to be in error. The relationship can be explained as follows. Most 100-kHz converter's sample and hold circuits are set to acquire the signal at about 80% of the sample window (10 µs) to allow 8 µs of settling time. Subtracting this from the scan time results in a sampling time of:

EQN. 3.02 Settling Time

$$T_{sample} = T_{scan} - T_{settle}$$
$$T_{sample} = 10\ \mu s - 8\ \mu s$$
$$T_{sample} = 2\ \mu s$$

In a typical 16-bit data acquisition system, the internal settling time (T_{int}) may be 6 µs. The external settling time may then be computed as follows:

EQN. 3.03 External Settling Time

$$T_{ext} = \sqrt{T_{settle}^2 - T_{int}^2}$$
$$T_{ext} = 5.29\ \mu s$$

For a 16-bit data acquisition system with 100 pF input capacitance, (C_{in}) and a multiplexer resistance (R_{mux}) of 100 Ω, the maximum external resistance is:

EQN. 3.04 External Resistance

$$R_{ext} = \frac{T_{ext}}{(C_{in})(l_n 2^{16}) - R_{mux}}$$

$$R_{ext} = 4{,}670\ \Omega$$

The simplified examples above do not include effects due to multiplexer charge injection or inductive reactance in the measurement wiring. In actual practice, the practical upper limit on source resistance is between 1.5 k and 2 kΩ.

Chapter 4

ELECTRICAL MEASUREMENTS

SAFETY CONCERNS

In the United States, electrical potentials more than 48 V are considered dangerous to human life. Although people have survived electrical shocks exceeding thousands of volts and lightning strikes carrying millions of watts of power, everyone should treat electricity with caution and respect. Voltage is not the only concern, however. Currents in the mA range can stop a normal heart under certain conditions. For example, NIOSH (National Institute for Occupational Safety and Health) estimates that as little as 20 mA at 110 Vac, the common utility voltage, can paralyze respiratory muscles.

All equipment and sources of voltage and current, such as the item under test, should not be powered when making connections between sensors, actuators or other components and the data acquisition system. This is especially critical when using current transformers and measuring high voltages and ac current.

MAKING VALID MEASUREMENTS

Accurate measurements depend on accurate equipment and robust connections. Poor connections and noise picked up through unshielded leads and improperly grounded circuits will certainly yield inaccurate data. Packaged sensors are typically designed to ensure accurate output signals with proper component placement, shielding, filtering, and bypassing power and ground I/O connections. Unprotected sensors, however, such as thermocouples and strain gages often are exposed to numerous types of noise signals that can couple into the sensors and leads and contaminate the desired signal. Those who install such sensors must follow industry-accepted practices to ensure that the sensors' output signals are reliable, unbiased, and free of noise and interference.

DC VOLTAGE

Instrumentation Level DC Voltage

Data acquisition systems usually contain integrated-circuit signal conditioners and multiplexers at their input terminals. These ICs typically can't handle more than about ±10 Vdc input, although some may tolerate as much as ±25 Vdc. One way around the limitation is to place a voltage divider on the system's input terminals that reduces the voltage to meet the IC's specifications. Some systems are manually switched to the proper range, while others select the range automatically.

In addition, after properly connecting the sensors, the item under test, and the data acquisition system, apply power first to the data acquisition system. Next energize the sensors and finally, the item under test. Some data acquisition systems can be damaged when voltages are applied to their inputs while unpowered.

High DC Voltage

Voltages greater than about 24 Vdc are considered high voltage in data acquisition system terms. Built-in or external voltage dividers and special signal conditioners are needed for dividing tens or hundreds of volts down to just 10 V or less to protect the instrument's input circuitry. One critical consideration for the safety of both the equipment and operators is to use high-voltage insulation in wires, terminals, connectors, leads, sensors, and other components to prevent leakage and arcing to low-potential terminals and nearby objects.

Low DC Voltage

Compared to high voltages, successfully measuring low-level dc signals depends upon appropriate wiring techniques between the signal source and the data acquisition device. Input amplifier stages in signal conditioning equipment cannot distinguish between the measured signal and a noise voltage coupled into lead wires. When measuring signals less than one volt, shielded or unshielded twisted pairs provide the best protection against noise voltage pickup. Best wiring practices call for shields to be grounded at one end only, preferably at the signal source.

AC VOLTAGE

Low AC Voltage

General-purpose data acquisition systems often measure low-level ac voltages on the secondary of loaded current transformers, step-down potential transformers, clamp-on current probes, current-sensing resistors, or sensors lacking galvanic isolation. All these low-level ac voltages must have a low common-mode voltage component with respect to earth ground or float with minimal ac leakage. Also, connect the voltage source to earth ground when necessary.

Outlet Level AC Voltage

In the U.S. and Canada, utility wiring supports a nominal 110 Vac, 60 Hz, single-phase voltage, supplied with three wires; hot, neutral, and ground. Europe typically uses 220 Vac, 50 Hz, and Japan supplies 100 Vac; 50 Hz in Eastern Japan and 60 Hz in Western Japan. In many factories, 220/230 to 460/480 Vac, single and three-phase voltage, Wye and Delta configurations are also distributed to electrical and electronic control equipment. All these voltages are lethal and leads should be connected or disconnected from the data acquisition system and the equipment under test with the power disconnected.

When connecting these voltage sources to data acquisition systems, problems often crop up in ground loops, which produce noise. Voltage drops appear between the neutral wire and ground for many reasons, some of which cannot be eliminated easily or even found. Accepted practices for shielding, grounding, and isolation should be followed rigorously. Refer to Chapter 10 for more detailed information concerning ground loops and interference.

High Voltage

Similar to dc voltage levels, data acquisition system inputs typically can't handle more than ±10 Vac peak. Higher ac voltages are attenuated and scaled with a signal conditioning stage before they are processed. Voltages up to 2,000 V, peak-to-peak, can be measured with a fully differential attenuator, calibrated to match the buffered differential inputs of a data acquisition system. Attenuators are typically voltage divider networks connected from the high side input terminal to a common ground.

For systems with a true differential input, however, neither the input's low nor high side can be connected directly to ground. Two matched resistor dividers attenuate the incoming signal, one on the high side and the other on the low side. Only the low side of the attenuator is grounded. For example, an accessory device is required to provide a 200:1 attenuation without affecting the amplifier's high input impedance. In a typical system, the two matched buffer amplifiers, A and B (Figure 4.01), on the front end of the data acquisition system draw almost identical

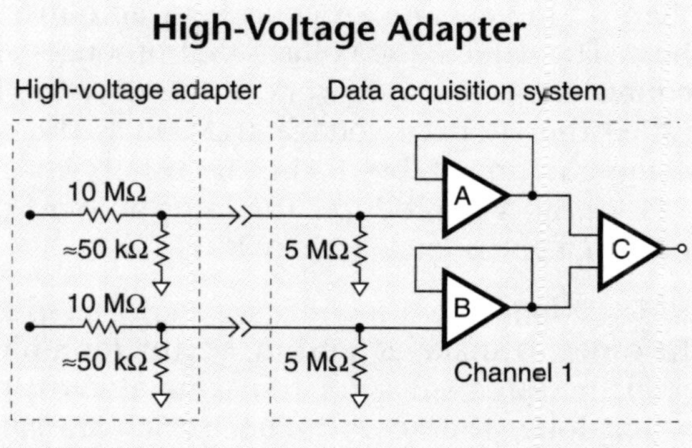

Fig. 4.01. A typical data acquisition system usually requires an attenuator at its input to accept voltages that exceed 10 V.

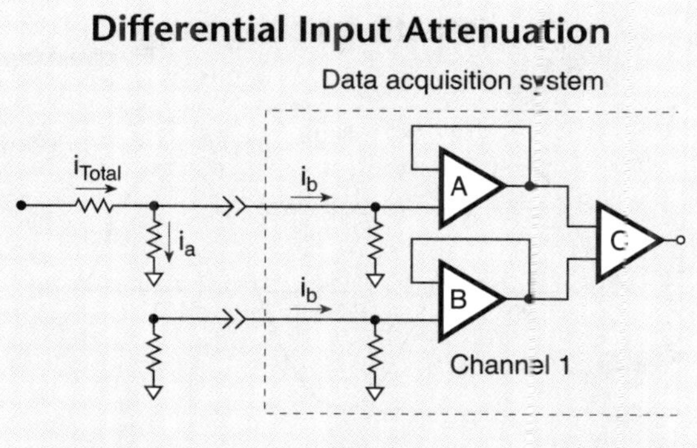

Fig. 4.02. A differential amplifier requires a resistor at both inputs when running single-ended, but only one resistor in a voltage-divider attenuator.

input bias current. Matched voltage dividers placed on the high and low-side inputs provide symmetrical current limiting from either input and let the quasi-differential input of the data acquisition system measure a lower voltage referenced to analog common. Because the outputs from amplifiers A and B are identical, they can be considered a common-mode voltage that the differential amplifier C rejects.

This scheme attenuates the high and low signals, and lets both ends float hundreds of volts above ground. For example, the high input can be 110 Vac while the low end has a 50-Vdc offset. After attenuation, the peak voltage of the high side signal is 0.77 V and the low side signal is 0.25 V, which are well within the readable range of the data acquisition system's differential inputs.

When a single-ended measurement is attempted by attenuating only the high signal and pulling the low signal to ground, two problems can develop. First, when the data acquisition system ground and the signal ground are different, tying them together forms a ground loop, which can result in excessive current and destroy the instrument or equipment. The second error component comes from the bias currents I_b flowing through the input pins of the two operational amplifiers, A and B shown in Figure 4.02. The bias current produces a voltage drop through the divider on the high side input, but the low side input has no matching drop. The difference appears at the input to differential amplifier C and produces an offset at its output.

Effective, Peak, Average, and RMS Voltage

Dc voltages and currents are relatively easy to measure and calculate with Ohm's law. For example, dc power is mathematically equivalent to the product of dc voltage and current and generates a unique amount of heat energy or performs a unique amount of work. By comparison, power calculations for ac voltage and current includes an additional factor, the cosine of the phase angle between them, called the power factor. When the voltage and current waves are in phase, the cosine of the angle is unity, so the product of voltage and current is a valid power value measured in watts. When they are out of phase, however, the power factor is less than unity, so the power generated is less than the simple product of voltage and current. In an extreme case, 90° for example, the cosine of the phase angle is zero, so the wattage is also zero. However, the power is then totally reac-

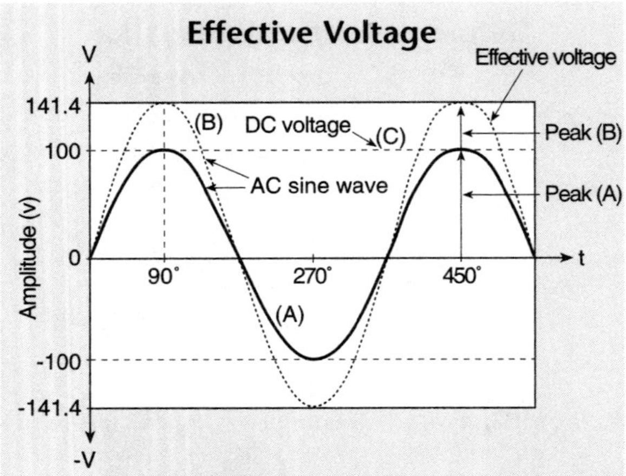

***Fig. 4.03.** The numerical value for the peak voltage of ac must be 1.414 times that of dc to produce the same heating effect.*

tive, does not generate heat (theoretically), and is simply labeled VA (volt-amperes).

EQN 4.01. DC Power

P = E x I
Where:
E and I are in phase
P = power, W
E = dc voltage, V
I = dc current, A

EQN 4.02. AC Power

P = E x I x Cos ϕ
Where:
P = power, W
E = ac voltage, V
I = ac current, A
ϕ = phase angle between voltage and current

Effective voltage

The ac voltage wave shape follows a sine function (ideally), and because it continually rises and falls over each cycle, the area under the curve for ac is less than for dc over the same (peak) amplitude and time period. Thus, 100 Vac peak generates less heat than 100 Vdc in the same load. (See Figure 4.03.) To compensate for this, peak ac voltages must be higher than dc to generate an equivalent amount of heat. The value of ac voltage that produces an equivalent amount of work as the dc value is called the effective voltage and equals the dc value multiplied by 1.414. In other words, 141.4 peak Vac produces the same amount of heat in a given load, as does 100 Vdc.

RMS voltage

In order to make ac and dc voltmeter scales read the same value for the same amount of work done, the ac voltmeter scales are calibrated in rms (root mean square) voltage values. When observing 100 Vac rms on the meter face, the voltage is actually 141.4 Vac peak, and the amount of work it does equals that of 100 Vdc (in the same load). Moreover, for ac, the maximum (peak) positive excursion and the maximum negative excursions are the same; therefore the peak-to-peak value in this example is twice 141.4 or 282.8 Vp-p.

The rms voltage value of a sine wave is equal to the effective value and is defined as the square root of the average of the squares of the instantaneous voltages measured over a given number of cycles.

EQN 4.03. Volts rms

When the shape of the ac signal is not a near-perfect sine wave, ordinary vane-type voltmeters will not indicate a true rms ac value. Special meters and some data acquisition systems, however, are designed to compensate for the shape discrepancies and are called true rms meters. They measure and indicate the true rms value of the ac signal, regardless of the distortion. The average value of the ac voltage is zero, because the positive half cycle magnitude equals the negative half cycle of the wave shape.

CURRENT

A Fundamental Problem

Voltage drops are relatively easy to measure

with most instruments because the difference in voltage appears between any two points in a circuit. The circuit does not need to be disturbed or changed (provided the measuring instrument's impedance is high compared with the impedance of the measured circuit) to make the measurement. By comparison, simple direct-connection current measurements are more troublesome, because the current appears within a loop and the loop must be opened to insert the ammeter. Moreover, the ammeter contains a characteristic resistance or impedance that most often changes the circuit parameters and must be compensated to obtain accurate measurements.

Some current measuring instruments, however, can couple inductively and have little or no affect on the circuit parameters. Such probes are designed for both ac and dc circuits.

Insertion Loss

Inserting an instrument or device in series with the circuit causes the total series impedance to change. In many applications, the added component or sensor requires some power to operate or drops some voltage resulting in what is called insertion loss. Therefore, some signal power may be consumed in making the measurement, which must be compensated to provide a more accurate measurement.

Common-Mode Limitations

Most differential amplifiers used for signal conditioners in data acquisition systems can sustain only a limited common mode voltage, typically about ±10 Vdc. However, more expensive amplifiers or signal conditioners are available for special applications that sustain common mode voltages exceeding the 10-Vdc limit.

One application, for example involves measuring the voltage drop across a calibrated shunt resistor in an alternator charging circuit. When the amplifier input is connected to the shunt resistor to measure mV scaled to charging current, the common mode voltage

Typical Data Acquisition Input

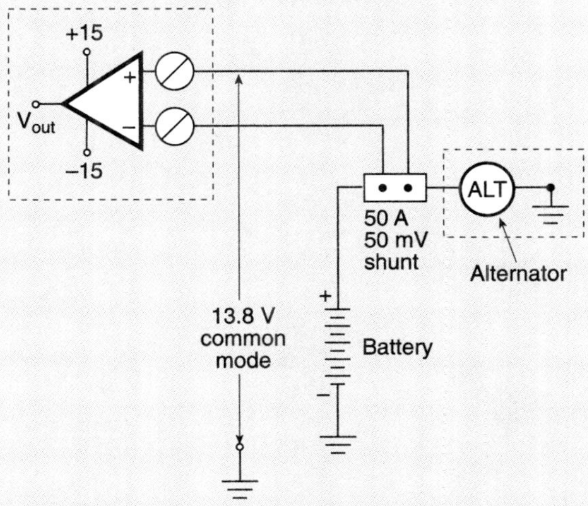

Fig. 4.04. *The battery charging current in an automobile electrical system is typically measured with a handheld diagnostic voltmeter across a shunt resistor, which is totally isolated from the battery voltage. However, when a data acquisition input amplifier is connected to the shunt, the input also must sustain the common mode voltage represented by the battery.*

seen by the amplifier input is the maximum battery voltage, nominally 13.8 Vdc, but may be as high as 18.5 Vdc during high-current charging. Therefore, the common-mode voltage rating for the data acquisition system input must be greater than 18.5 Vdc.

Shunts

Data acquisition instruments usually measure high-level dc currents (in the ampere range) as a 50- or 100-mV voltage drop across a calibrated shunt resistor. Figure 4.05 shows a special application for differential measurements where the three shunts have one end placed in the common side of a three-motor circuit. This arrangement provides a common ground that can be connected to an inexpensive monitoring unit with a non-isolated, low-voltage input. Unfortunately, not all circuits that need more than one shunt will have one end of each shunt

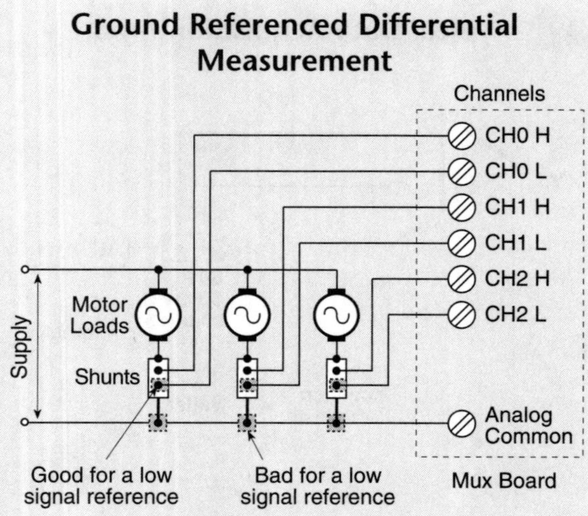

Fig. 4.05. High dc currents are often measured with shunts using differential inputs sharing a common ground.

conveniently connected to a common point. In these instances, more expensive, isolated analog input amplifiers are recommended.

Data acquisition systems also measure low-level dc currents with shunts. Many sensors output a standard 4 to 20 mA current, which is linearly related to the quantity being measured. The signal current passes through the shunt resistor, as shown in Figure 4.06, and the voltage drop across the shunt provides the input for the ADC.

Ohms law defines the nature of this simple circuit where the product of resistance (R) and current (I) yields the voltage: V = IR. Consequently, when the data acquisition system requires a 10-V input for a 20-mA full-scale reading, a 500-Ω resistor does the job:

EQN. 4.04. Shunt Resistor

$$R = V/I = 10\ V/0.02\ A = 500\ \Omega.$$

The current loop provides higher noise immunity and accuracy than does a voltage source

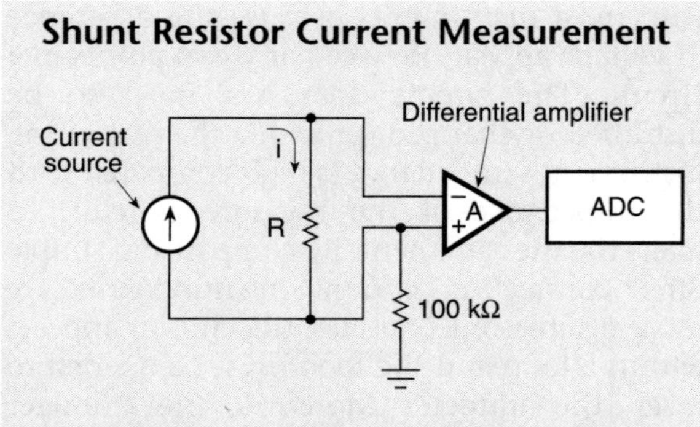

Fig. 4.06. Sensors with 4 to 20 mA outputs are typically placed in series with a resistor to develop a voltage. The resistor is then connected to the input of a differential amplifier to measure the voltage drop.

when measuring with long leads of relatively high resistance. In this case, the lead resistance is part of the 500 Ω, so in a simple series circuit, the shunt resistor is often adjustable in order to calibrate the loop to exactly 20 mA full scale.

The accuracy and stability of the measured voltage across the shunt are only as accurate and stable as the shunt resistor itself. Resistor accuracies are commonly 5, 1, 0.5, 0.1 and 0.01% and come with a temperature coefficient that specifies their stability over a specific temperature range. In addition, a 0.1% resistor usually has a lower temperature coefficient and better long-term stability than a resistor with lower accuracy because of its construction, which is usually wire wound or metal film vs. carbon film or carbon composition.

Current Transformers

High ac currents can be measured with shunts under special circumstances, but most often such direct connections to ac lines are extremely dangerous to humans. One way to get around the hazard is to use current transformers (CT) that isolate ac line voltages and reduce input current by a specified ratio. (See Figure 4.07.) A 500:5 CT, for example,

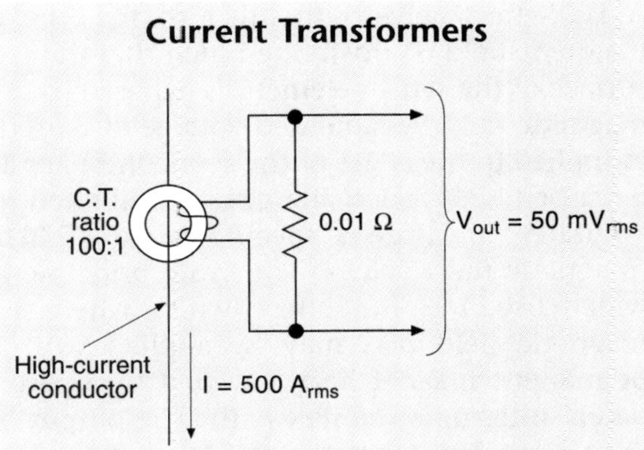

Fig. 4.07. Current transformers are a convenient means to measure high ac currents. The voltage dropped across the resistor is proportional to the current through the transformer.

has a 100:1 ratio and generates a secondary current of 5 A with 500 A in the primary. When a low value load resistor such as 0.01 Ω connects to the output of the current transformer, the full load secondary current produces 50 mV_{rms}, which can be easily read with the data acquisition system's analog input. Although this output voltage seems low, higher resistance values will drive the CT out of its calibrated range and reduce the accuracy of the measurement. Moreover, a CT rated at 2 VA can handle only a maximum usable load resistance (including lead wires) of 0.08 Ω.

For safety, always make certain that the current loop is closed before applying power. An open-circuit CT can generate many thousands of volts at its terminals with lethal current levels. For example, consider a 5,000:5-A current transformer with a turns ration of 1,000 to 1, and 10 Vac output signal developed across a calibrated low impedance load. The open-circuit secondary voltage is then 1,000 X 10 = 10,000 Vac, a very dangerous level to be connecting or disconnecting to a data acquisition system shunt resistor.

Hall-Effect Sensors

In 1879, Dr. Edwin Hall discovered the fundamental operating principle of the sensor named after him. The basic Hall-effect sensor operates under the influence of three parameters; current, voltage, and a magnetic field. Hall passed a constant current through a gold foil conductor while exposing the surface of the conductor to a magnetic field. A voltage drop developed across the opposite sides of the conductor with a magnitude dependent on the property of the sensor material (Hall coefficient) and proportional to the intensity of the magnetic field. (See Figure 4.08.)

A modern Hall device replaces the original conductor with a thin sheet of semiconductor material. The output voltage appears between two opposite edges of the sheet when the flat surface is exposed to a magnetic field, and a calibrated current is injected across the other two opposite edges.

The Hall-effect sensor is ideal for measuring magnetic fields or ac and dc power. For power applications, it becomes a multiplying device. It automatically multiplies the voltage and current to yield power and includes the phase

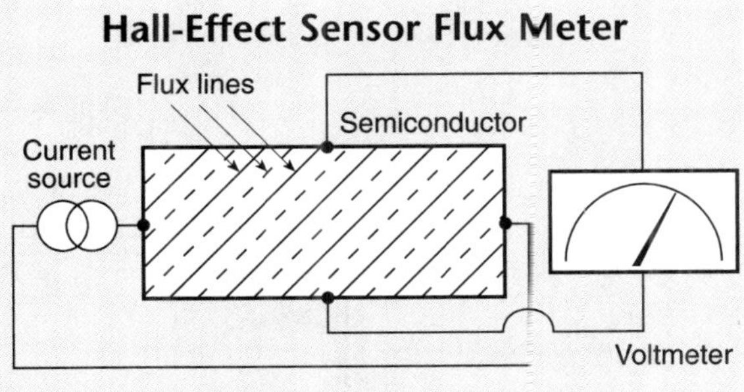

Fig. 4.08. When either the voltage or current port is regulated and calibrated, the Hall-effect device can be used as a magnetic flux sensor. The unregulated port provides an output signal proportional to the magnitude of unknown flux density. However, a Hall-effect proximity sensor or switch is a more widely used component based on this principle.

angle factor as an integral property. An electromagnetic coil generates the Beta field with its core perpendicular to the surface of the thin Hall element inside a shielded package. The coil connects in series with the load to sense load current, and the load voltage connects across the semiconductor element as shown in Figure 4.09. The resulting current vector drives a readout device calibrated in watts or becomes a control signal to a power controller in a closed-loop circuit. Because the Hall-element material is temperature sensitive, a packaged Hall-effect wattmeter contains temperature-compensating circuits. The defining equation for the output is the product of the vectors:

EQN: 4.05. Hall-Effect Sensor Power

$P = E \times I \times \cos(\phi)$

Where:

P = power in the load, W

E = load voltage, V

I = load current, A

ϕ = phase angle between load voltage and load current

A Hall-effect sensor intended to measure a magnetic field is constructed such that the flat surface of the sensor element is exposed to the magnetic field. A calibrated voltage or current is applied to the edges of the semiconductor as described above, and the unit is calibrated to accurately measure a specific range of Beta (magnetic flux) values. The same principle is widely used in a Hall-effect switch, where the magnetic field bias may be applied with a permanent magnet and a slotted vane alternately interrupts the flux path. The output of the sensor switches between a low voltage and a high voltage as the flux path is interrupted. Such a device had been used as a switch for many years in numerous automobile distributors to replace the ignition points.

RESISTANCE

Basic Measurements

Ordinary ohmmeters measure resistance indirectly by injecting a small current into the resistor under test and measuring the resulting voltage drop across its terminals. Precision

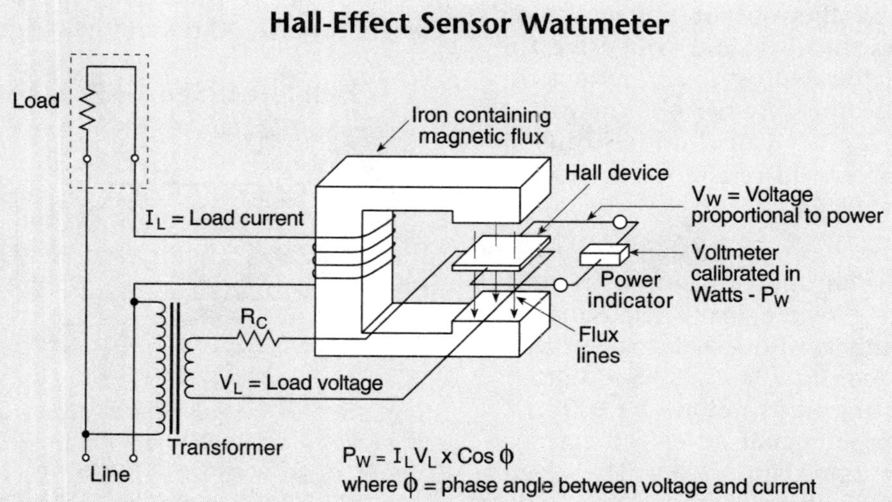

Fig. 4.09. A Hall-Effect sensor is an ideal wattmeter for measuring ac power. It inherently includes the phase angle between the voltage and current vectors as a multiplying factor to convert VA to W.

reference resistors called multipliers scale analog meters to read in the middle portion of its face, while digital ohmmeters scale either manually or automatically to properly position the decimal point. Digital multimeters (DMMs) similarly apply a dc voltage across an unknown resistor and a precision shunt resistor. The test current generates a voltage drop across the shunt resistor, which is read by the ADC to output a digital value of the unknown resistor. Data acquisition systems, however, do not typically read ohms directly; they measure voltages and voltage ratios across known and unknown resistance dividers and compute equivalent resistor values.

Voltage Divider Method

Resistance voltage divider circuits can use either single-ended or differential-input amplifiers. (Refer to Figure 4.10.) The known reference resistor should have a value that is close to the unknown value, and the source voltage should be low enough to ensure that the current is within acceptable limits such as 1.0 mA or less. For example, when measuring an unknown resistor near 1,000 Ω, select a 1,000 Ω reference resistor and a supply of 2.0 Vdc. This produces a current of 2/2,000 = 1 mA. A 1,000-Ω reference resistor keeps the current relatively low and satisfactorily measures unknown values down to the 100-Ω range. The unknown resistor value is calculated from:

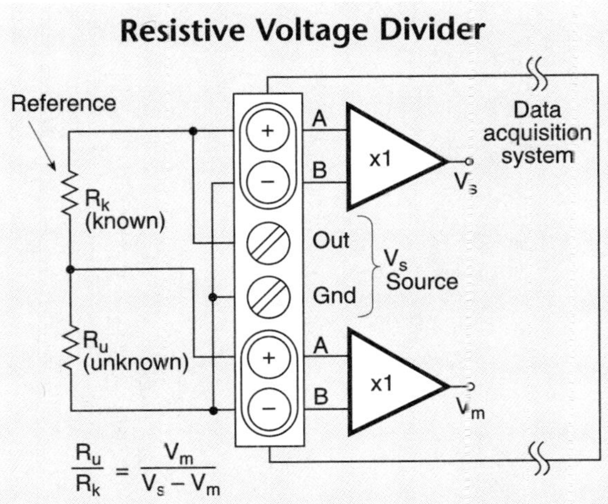

Fig. 4.10. Resistive voltage dividers work with single-ended or differential amplifiers and a reference resistor whose value is near the unknown resistor's value.

EQN. 4.06. Voltage Divider Resistor

$R_u = (R_k)(V_m)/(V_s - V_m)$

Where:

V_m = measured input voltage, V

V_s = source voltage channel, V

R_k = known resistance, Ω

R_u = unknown resistance, Ω

For example:

Where:

$V_s = 2$ V

$V_m = 1.10$ V

$R_k = 1{,}000\ \Omega$

R_u = unknown resistor, Ω

Then:

$R_u = (1\text{ k})(1.10\text{ V})/(2.0 - 1.10)$

$R_u = (1{,}100)/(0.90) = 1{,}222.2\ \Omega$

Constant Current/Voltage Drop

Current source/voltage drop methods to measure unknown resistances use both two and four-wire circuits. Two-wire circuits should be avoided for measuring very low to moderately low resistances because the lead wires and connections can become part of the total resistance (which is added to the unknown value).

Small Deviations

Resistance measurement errors come from a number of sources. The most common errors in low resistance circuits for simple constant current/voltage drop methods using two wires come from the lead wire resistance and connections added to the unknown resistance. Keeping the lead wires short and using four-wire circuits such as Kelvin connections can minimize these errors. Errors in high resistance measurements can come from the shunt resistance of insulation, such as in wiring and connector bodies. When using a current source and measuring the voltage drop, use high input impedance measuring devices, whether they are voltmeters or signal conditioner inputs. For extremely high resistor measurements, use a voltage source and measure the current using short lead wires and robust connections.

WHEATSTONE BRIDGE

Basic Principle

Wheatstone bridges often measure precise resistances and extremely small resistance changes. A typical bridge circuit is arranged in a symmetrical pattern of equal-value series-parallel resistors across a power source as shown in Figure 4.11. When all resistors are identical, the voltage drops across them are equal, and the voltage between nodes A and B is zero. When one or more resistors are not equal, which is usually the resistor under test, the bridge unbalances and a null meter placed between the nodes measures the amount of unbalance. An adjustable resistor or a potentiometer can be placed in one arm of the bridge to manually rebalance the circuit. When the potentiometer drives the meter back to zero, its value then equals that of the unknown resistor. The potentiometer comes with a precision, calibrated readout device that indicates its resistance, which is also the value of the unknown resistor.

Anderson Loop Alternative

The basic Anderson loop is often compared to a passive bridge circuit, but it provides higher accuracy measurements with less excitation

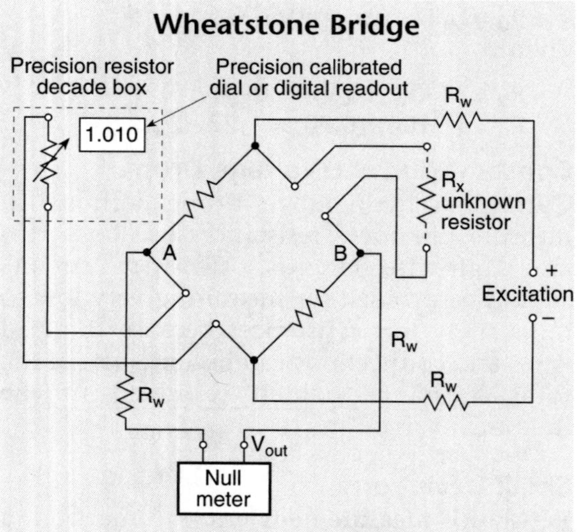

Fig. 4.11. An unknown resistor placed in one arm of a Wheatstone bridge typically unbalances the circuit symmetry and produces an offset voltage across nodes A and B. But adjusting a precision, variable resistor in the opposite arm brings the circuit back into balance when it equals the unknown resistor value, which may be read on a calibrated dial.

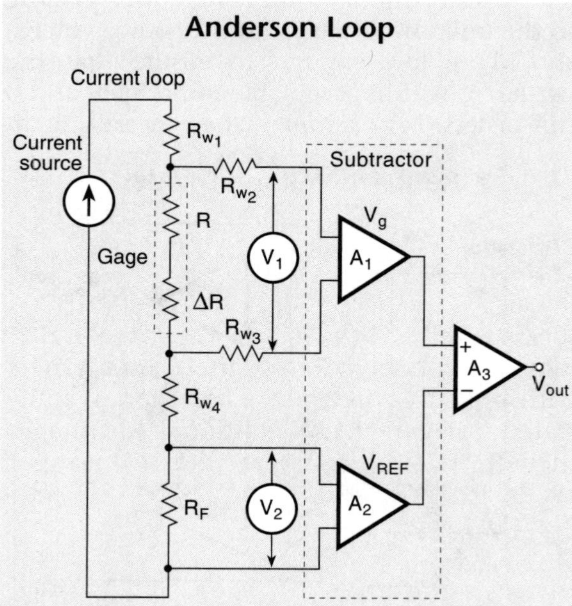

Fig. 4.12. The Anderson loop uses a unique subtractor circuit, which eliminates errors arising from lead wire resistance. It measures only the sensing element's resistance change.

voltage. Furthermore, it contains an amplifier, which puts it in the class of an active device.

The basic Anderson loop is a series circuit composed of one sensing element in series with one reference element powered by a constant-current source. A dual-differential operational amplifier with high temperature stability and high CMRR, called a subtractor, delivers the output response from the sensing element. (Refer to Figure 4.12.) The advantage of this topology lies in its ability to ignore changes in lead wire resistance and to amplify only the output changes of the sensing element.

The output response V_{out} of a basic strain-gage circuit is defined by:

EQN. 4.07. Anderson Loop Output

$$V_{out} = A_1V_1 - A_2V_2$$
$$V_{out} = I_r\,(\Delta R_g)$$

Where:

A_1 = amplification factor of stage 1
A_2 = amplification factor of stage 2
V_1 = voltage across sensing strain gage, V
V_2 = voltage across reference strain gage, V
I_r = constant current, A
ΔR_g = strain gage resistance change, Ω

The circuit is not limited to only one sensor; it can be expanded to contain numerous sensors in series such as a strain rosette. It may be connected to signal conditioners with only six wires plus a common as illustrated in Figure 4.13. Additional advantages include an output voltage that is linear for each gage response, it requires only 1/4 the power needed by the conventional bridge circuit, and the loop output voltage is twice that of a conventional strain gage. The fact that the output signal of the Anderson loop is twice that of a strain gage Wheatstone bridge circuit accounts for a 6 dB increase in signal-to-noise ratio.

The output responses of a multiple strain gage arrangement can be expressed as:

EQN 4.08. Multiple Strain Gage Output

$$V1_{out} - V_{ref} = I\Delta R1$$
$$V2_{out} - V_{ref} = I\Delta R2$$
$$V3_{out} - V_{ref} = I\Delta R3$$
$$V4_{out} - V_{ref} = I\Delta R4$$

Where:

$R_{ref} = R_n$, nominal gage resistance
V_{nout} = output voltage, V
I = constant current, A
ΔR_n = the change in resistance of the nth strain gage, Ω
V_{ref} = reference voltage, V

Anderson Loop for Multiple strain Gages

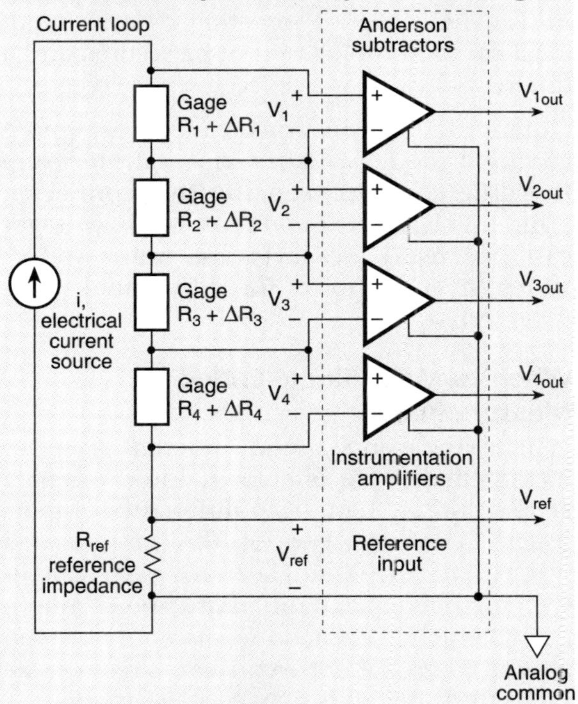

Fig. 4.13. The Anderson loop may be connected to multiple sensors in series such as in this strain rosette.

The significance of the output signal voltage, $I\Delta R_n$, is that it is principally a function of the change in resistance or impedance of the sensing element, not it's absolute terminal resistance.

SINGLE-ENDED AND DIFFERENTIAL MEASUREMENTS

Basic Principles

Data acquisition systems provide for both single-ended and differential-input connections. The basic difference between the two is the choice of the common connection for analog voltage inputs. Single-ended multichannel measurements require that all voltages be referenced to the same, well-chosen common node to prevent certain types of measurement errors. Sometimes, however, an ideal common point cannot be secured and a differential input is needed.

Differential connections cancel or ignore common-mode voltages and can measure the difference between the two connected points. The rejected common mode voltages can be steady dc levels or noise spikes. When given a choice, the differential input is preferred.

When to Make Single-Ended Measurements

When single-ended and differential input measurements are offered in the same data acquisition system, the differential inputs usually consume two of the single-ended inputs. That is, 16 single-ended input amplifier channels equal 8 differential-input amplifier channels. When noise susceptibility and ground loops are not a concern, single-ended input measurements may be satisfactory.

When to Make Differential Measurements

Differential measurements call for a differential amplifier to measure the difference voltage between the two input signal leads. Differential amplifiers are preferred because they provide greater noise rejection than single-ended amplifiers. Furthermore, some sensors, particularly strain gages, require measuring the voltage difference between the two signal leads. Figure 4.14 depicts a differential amplifier configured for a thermocouple. Although the amplifier measures the voltage difference between the two inputs, a path to ground is required on at least one input of the amplifiers. The 10 kΩ resistor between the low side and ground provides a path for amplifier bias current. If the resistor is absent or too high in value, the bias current drives the input to one of the power supply rail values, saturating the amplifier, which exceeds its usual ±10-V common-mode range.

The common-mode voltage generated by the bias current, i_b, is:

Differential Thermocouple Measurement

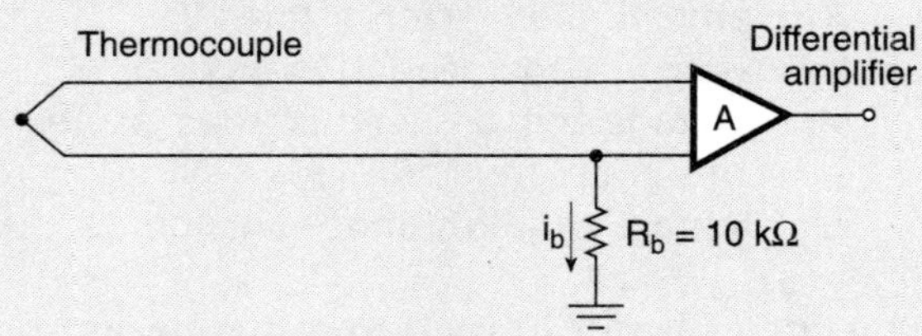

Fig. 4.14. The resistor is necessary for at least one input to provide the path to ground for amplifier bias currents.

EQN. 4.09. Common Mode Voltage

$$V_{cm} = 2i_bR_b$$

Where:

i_b = bias current, A

R_b = bias resistor, Ω

The minimum value for R_b may be as low as a short circuit (on one side only), which reduces to a single-ended input configuration. Otherwise, under worse-case conditions, the voltage error in percent of the applied voltage due to source resistance and R_b is:

EQN. 4.10. Voltage Error

$$\Delta E(\%) = 100\left(\frac{R_S}{2R_b}\right)$$

Where:

ΔE = error, %

R_S = source resistance, Ω

R_b = bias resistor, Ω

This applies to R_b for a balanced circuit where bias resistors are on both inputs. Values for R_b typically range from 10 kΩ to 10 MΩ. If both equations cannot be satisfied simultaneously, the source resistance is too high to operate within specifications.

Some amplifier inputs come with built-in bias resistors for differential measurements, while users supply the bias resistors for extremely high-impedance inputs. However, an incorrect user-supplied bias resistor might dramatically lower the input impedance in non-isolated systems and load the circuits being measured. Other data acquisition systems provide a suitable range of resistors that can be switched to analog common (preventing users from selecting an incorrect resistor). Alternatively, users can disconnect all resistors when a dc path already exists at the sensor for bias current to flow from the amplifier inputs to analog common.

Consider the strain gage in Figure 4.15. When the excitation voltage is referenced to the same common as the differential amplifier, bias resistors are not required because the strain gage itself provides a current path to ground for the instrumentation amplifier. However, when the instrumentation amplifier is floating or referenced to a different common, at least one bias resistor is required. A balanced termination with a 100 kΩ resistor can also be used.

Common-mode rejection ratio, CMRR, is a critical amplifier specification and a measure of quality for instruments making differential

Differential Strain Gage Measurement

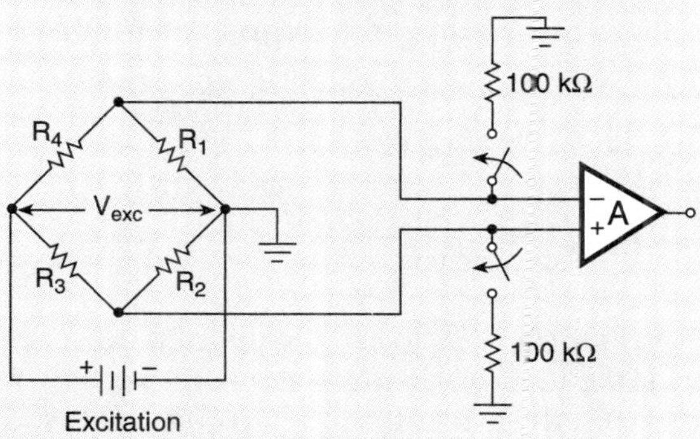

Fig. 4.15. When the strain gage reference voltage and amplifier share the same common terminal, switched bias resistors are not necessary.

measurements. A large CMRR is necessary to accurately measure a small differential signal on a large common mode voltage. The CMRR specifies the maximum effect of the common mode voltage on a differential measurement. Figure 4.16 shows the output signal free of the common mode input noise after passing through the differential amplifier. The two noise spikes entering the amplifier are equal in phase and amplitude, so the difference amplifier cancels them.

Differential Mode Noise Rejection

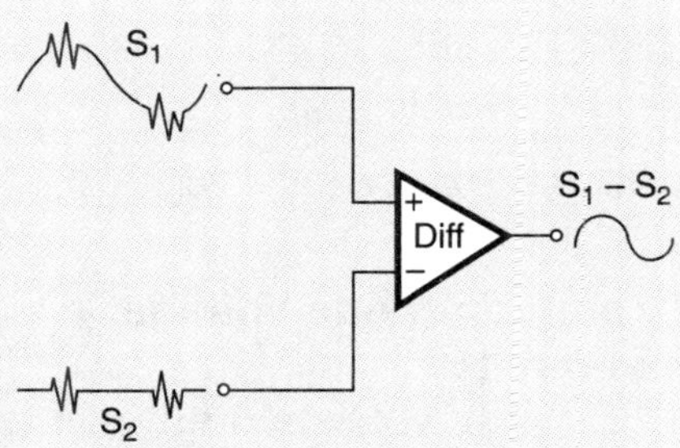

Fig. 4.16. The differential amplifier rejects voltages that are common to both lines.

An amplifier's CMRR is typically specified in decibels, dB, a ratio of two voltages that may differ by a large magnitude, and determined by:

EQN. 4.11. Common Mode Rejection Ratio

$$dB = 20\log(V_1/V_2)$$

Where:

V_1 = larger voltage, V

V_2 = smaller voltage, V

For example, a ratio of 10 equals 20 dB, a ratio of 100 equals 40 dB and each additional factor of 10 adds another 20 dB to the sum. Furthermore, an amplifier with a specification of 100 dB corresponds to a ratio of 100,000, so an instrument with a CMRR of 100 dB measuring a signal on a common-mode voltage of 1 volt can have an error of 0.01 mV. If the common-mode voltage is larger, or if greater accuracy is needed, an amplifier with a CMRR of at least 120 dB would be needed, a ratio of one million to one.

Calculate the CMRR for a differential amplifier given in a data sheet:

Where A_{diff} = differential gain = 10

A_{cm} = common mode gain = 0.001553

EQN. 4.12. Common Mode Rejection Ratio

$$CMRR = A_{diff}/A_{cm}$$

$$CMRR = 10/0.001553$$

$$CMRR = 6,439$$

Find CMRR in dB:

EQN. 4.13. CMRR Example

$$CMRR_{dB} = 20 \log 6439 = 20(3.820) = 76.4 \text{ dB}$$

High Common-Mode Voltage Measurements

Often, a small voltage riding on a much larger voltage must be measured. For example, when a thermocouple sits on one terminal of a battery, the signal conditioner input must be able to measure the millivolt output of the thermocouple while suppressing the battery voltage. When the common-mode voltage is less than 15 V, an instrument amplifier with differential inputs can read the thermocouple voltage while ignoring the battery voltage. When the common-mode voltage is higher than 15 V, however, an isolation amplifier is usually required. (Also refer to Figure 4.04.)

Although isolation does not protect amplifiers against excessive normal-mode input voltage (voltage across a pair of inputs), it does protect against excessive common-mode voltage. Isolation eliminates a potentially large and damaging current that would flow from the signal source due to the common-mode voltage.

Several isolation methods inherently include high common-mode voltage rejection. Each channel can have an isolation amplifier, or a group of channels (not isolated from each other) can be multiplexed and digitized by an ADC before the digital data are isolated. Isolation barriers can be optical, magnetic, or capacitive. The most common types are optocouplers composed of infrared (IR) light-emitting diodes (LEDs) detected by photodiodes on the opposite side of a high-voltage, quartz barrier. Optocouplers can transmit numerical data in serial pulse trains or sine waves and pulse trains whose frequency or pulse width can be modulated with analog signals. Varying LED current, as shown in Fig. 4.17, can also transmit analog information. Magnetic barriers using transformers and capacitive barriers are generally internal and used in monolithic or hybrid isolation amplifiers.

In frequency-coupled isolation, a high-frequency carrier signal is inductively or capacitively coupled across the isolation barrier. The signal is modulated on the input side and demodulated on the output side to reproduce the original input signal.

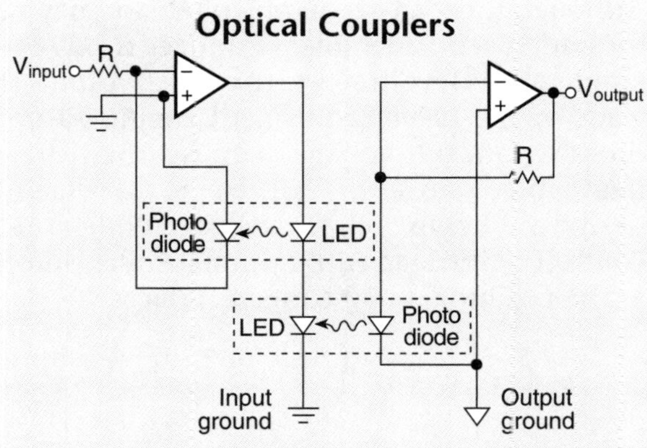

Fig. 4.17. When making high common-mode voltage measurements, optical couplers provide a convenient and safe means for connecting the voltage to data acquisition systems.

Isolated ADCs and their accompanying signal conditioning circuits float. The ADC converts the input signal to a digital signal, and the interface for transferring the digital code is digitally isolated. (See Chapter 6.)

KELVIN CONNECTIONS

Basic Principle

Wheatstone bridge circuits are accurate, simple, sensitive, and convenient ohmmeters to operate when the bridge, the unknown resistance, and the readout are all located near one another. But when the excitation and regulation circuit are located several meters away from the bridge circuit, the lead resistance, R_L, may cause an error in the excitation voltage.

One way around the problem is to use a Kelvin connection, a method that connects the excitation and regulator circuits to the bridge with four wires. Two wires carry the excitation current, and the other two, the sense lines, measure the excitation voltage at the bridge and carry no current. Because the signal conditioning circuit has high input impedance, little current flows in the input leads, so their resistance does not introduce significant errors. (See Figure 4.18.)

Examples of Kelvin-Type Measurements

An example of the Kelvin circuit illustrates an ohmmeter composed of a current source and a voltmeter. (See Figure 4.19.) The voltmeter measures the voltage dropped by the load resistor so the value of the resistor is simply V/I. The accuracy of the calculated resistance in this configuration is directly proportional to the accuracy of the ammeter and voltmeter. The resistance is calculated with Ohms law:

EQN: 4.14. Kelvin Connection Resistance

$R_U = V/I$

Where:

R_U = unknown resistance, Ω

V = voltmeter reading, V

I = ammeter reading, A

R_W = lead resistance

V_S = voltage source

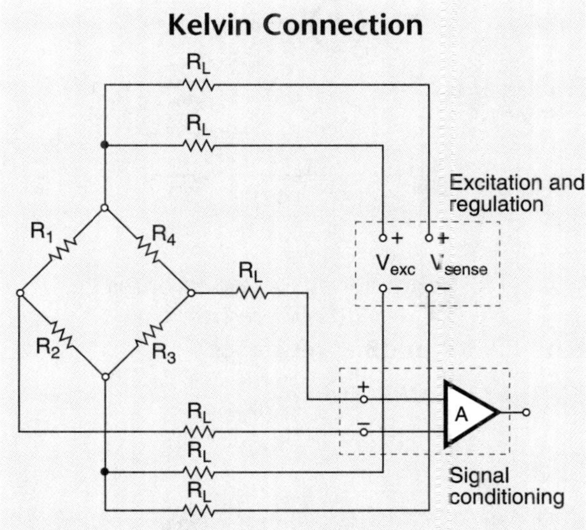

Fig. 4.18. The Kelvin connection eliminates lead resistance from altering the excitation voltage by using a second pair of wires. The measuring device has relatively high input impedance so it does not draw current and develop a voltage drop across its leads that would add an error.

Common Ohmmeter Circuits

Basic Ohmmeter Circuit

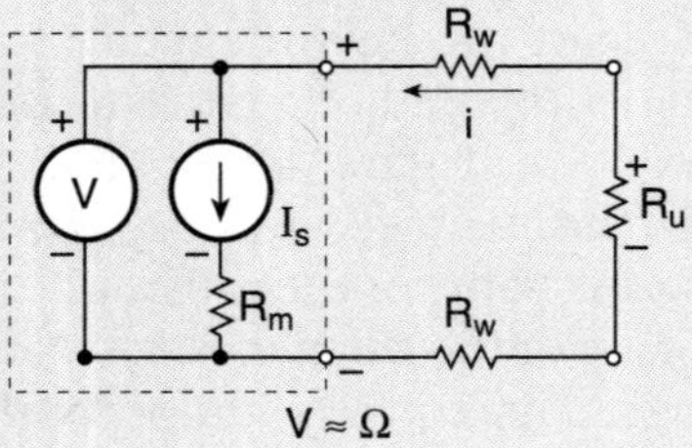

Kelvin Connection: Ohmmeter Circuit

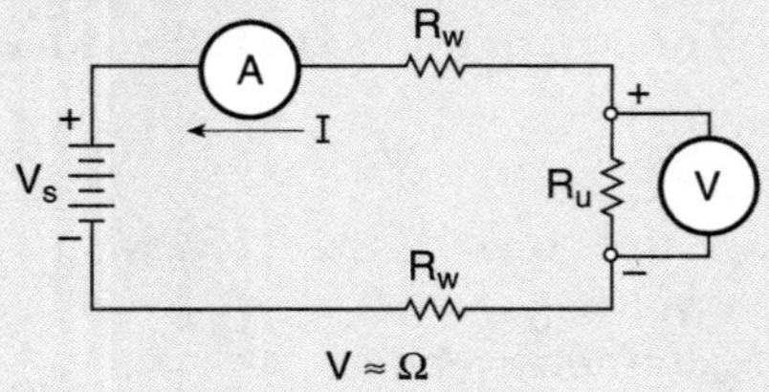

Kelvin Connection for Remote Voltmeter

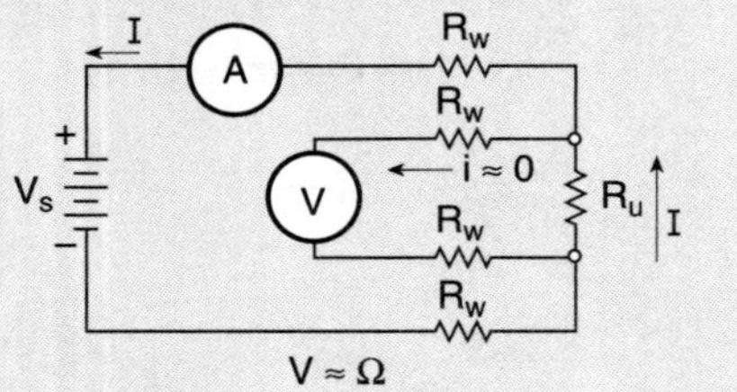

Voltmeter calibrated in ohms
R_m = Multiplier resistors
R_W = Lead resistance
R_u = Unknown resistor
V_S = Voltage source
I_S = Current source

Fig. 4.19. *The basic ohmmeter circuit contains a current source and multiplier resistors so the voltmeter scale may be calibrated in Ω. The Kelvin connection uses more than one set of leads to eliminate the lead resistance as a variable in the equation for both local and remote measurements.*

Although the load or unknown resistor may be located a considerable distance from the ammeter and voltmeter, the long voltmeter leads in the Kelvin equivalent circuit do not drop significant voltage because they draw negligible current. Moreover, specially designed Kelvin clips are available that connect to the load with minimal voltage drop at the jaw hinges and contact surfaces.

Chapter 5

FUNDAMENTAL SIGNAL CONDITIONING

AMPLIFICATION

Data Acquisition Front Ends

Data acquisition systems differ from single or dual-channel instruments in several ways. They can measure and store data collected from hundreds of channels simultaneously. However, the majority of systems contain from eight to 32 channels, typically in multiples of eight. By comparison, a simple voltmeter that can select a measurement among several different ranges can be considered a data acquisition system, but the need to manually change voltage ranges and a lack of data storage hobbles its usefulness.

Figure 5.01 illustrates a simple data acquisition system consisting of a switching network (multiplexer) and an analog-to-digital converter (ADC), both of which were discussed in Chapters 2 and 3. The main subject of this chapter, the instrumentation amplifier (IA), is placed between the multiplexer and ADC. The individual circuit blocks each have unique capabilities and limitations, which together define the system performance.

Data Acquisition Block Diagram

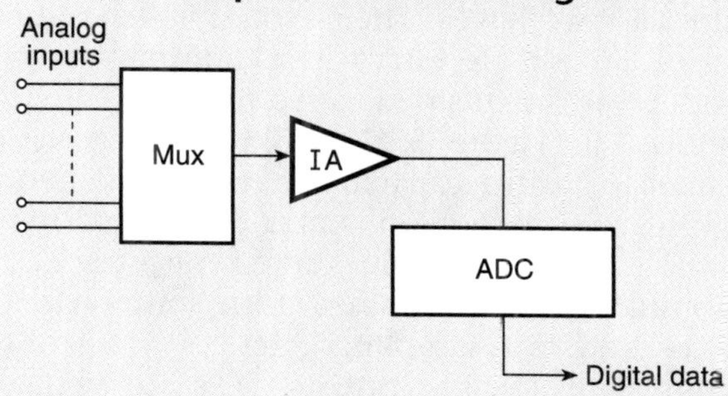

Fig. 5.01. *A simple data acquisition system is composed of a multiplexed input stage, followed by an instrumentation amplifier that feeds one accurate and relatively expensive ADC. This arrangement saves the cost of multiple ADCs.*

The ADC is the last in a series of stages between the analog domain and the digitized signal path. In any sampled-data system, such as a multiplexed data acquisition system, a sample-and-hold stage preceding the ADC is necessary. The ADC cannot digitize a time-varying voltage to the full resolution of the ADC unless the voltage changes relatively slowly with respect to sample rate. Some ADCs have internal sample-and-hold circuits or use architectures that emulate the function of the sample-and-hold stage. The discussion that follows assumes that the ADC block includes a suitable sample-and-hold circuit (either internal or external to the chip) to stabilize the input signal during the conversion period.

The primary parameters concerning ADCs in data acquisition systems are resolution and speed. Data acquisition ADCs typically run from 20 kHz to 1 MHz with resolutions of 16 to 24 bits, and have one of two types of inputs, unipolar or bipolar. The unipolar-type typically ranges from 0 V to a positive or negative voltage such as 5 V. The bipolar-type typically ranges from a negative voltage to a positive voltage of the same magnitude. Many data acquisition systems can read bipolar or unipolar voltages to the full resolution of the

ADC, which requires a level-shifting stage to let bipolar signals use unipolar ADC inputs and vice versa. For example, a typical 16-bit, 100-kHz ADC has an input range of –5 V to +5 V and a full-scale count of 65,536. Zero volts corresponds to a nominal 32,768 count. If the number 65,536 divides the 10-V range, the quotient is an LSB (least significant bit) magnitude of 153 µv.

Multiplexing through high source impedances does not work well. The reason that low source impedance is necessary in a multiplexed system is easily explained with a simple RC circuit shown in Figure 5.02. Multiplexers have a small parasitic capacitance from all signal inputs and outputs to analog common. These small capacitance values affect measurement accuracy when combined with source resistance and fast sampling rates. A simple RC equivalent circuit consists of a dc voltage source with a series resistance, a switch, and a capacitor. When the switch closes at T = 0, the voltage source charges the capacitor through the resistance. When charging 100 pF through 10 kΩ, the RC time constant is 1 µs. In a 10-µs-time interval (of which 2 µs is available for settling time), the capacitor only charges to 86% of the value of the signal, which introduces a major error. But a 1 kΩ resistor lets the capacitor easily charge to an accurate value in 20 time constants.

Figure 5.03A shows how system input impedance and the transducer's source impedance combine to form a voltage divider, which reduces the voltage read by the ADC. The input impedance of most input channels is 1 MΩ or more, so it's usually not a problem when the source impedance is low. However, some transducers (piezoelectric, for example) have high source impedance and should be used with a special charge amplifier. In addition, multiplexing can greatly reduce a data acquisition system's effective input impedance. (See also Chapter 3, Multiplexing, Input and Source Impedance.) The charge injection effects are shown in Figure 5.03B.

Parasitic RC Time Constant

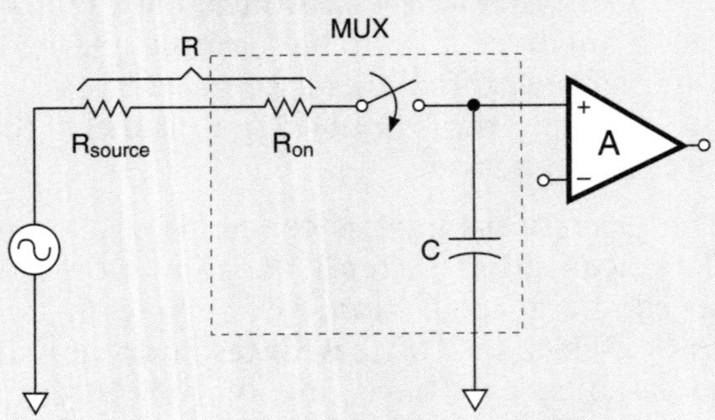

Fig. 5.02. The source resistance should be as low as possible to minimize the time constant of the MUX's parasitic capacitance C and series resistance R. An excessively long time constant can adversely affect the circuit's measurement accuracy.

Input and Source Impedance

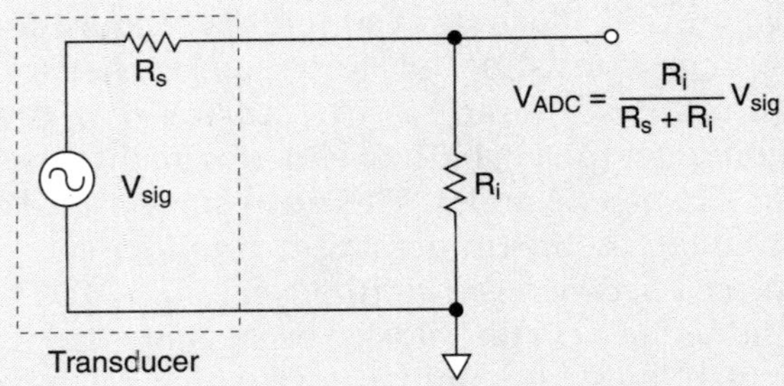

Fig. 5.03A. The sensor's source impedance R_S should be relatively small to increase the voltage divider drop across R_i, the amplifier's input. This can substantially improve the signal-to-noise ratio for mV-range sensor signals.

Operational Amplifiers

Many sensors develop extremely low-level output signals. The signals are usually too small for applying directly to low-gain, multi-

MUX Charge Injection Effects

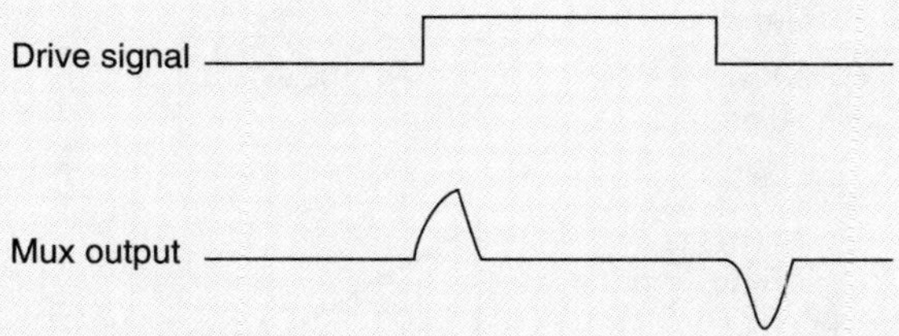

Fig. 5.03B. Analog-switching devices can produce spikes in the MUX output during level transitions in the drive signal. This is called charge-injection effect and can be minimized with low source impedance.

plexed data acquisition system inputs, so some amplification is necessary. Two common examples of low-level sensors are thermocouples and strain-gage bridges that typically deliver full-scale outputs of less than 50 mV.

Most data acquisition systems use a number of different types of circuits to amplify the signal before processing. Modern analog circuits intended for these data acquisition systems comprise basic integrated operational amplifiers, which are configured easily to amplify or buffer signals. Integrated operational amplifiers contain many circuit components, but are typically portrayed on schematic diagrams as a simple logical functional block. A few external resistors and capacitors determine how they function in the system. Their extreme versatility makes them the universal analog building block for signal conditioning.

Most operational amplifier stages are called inverting or non-inverting. (See Figure 5.04.) A simple equation relating to each configuration provides the idealized circuit gains as a function of the input and feedback resistors and capacitors. Also, special cases of each configuration make up the rest of the fundamental building blocks, namely the unity-gain follower and the difference amplifier.

Inverting Amplifier Stages

The inverter stage is the most basic operational amplifier configuration. It simply accepts an input signal referenced to common, amplifies it, and inverts the polarity at the output terminals. The open-loop gain of a typical

Operational Amplifiers

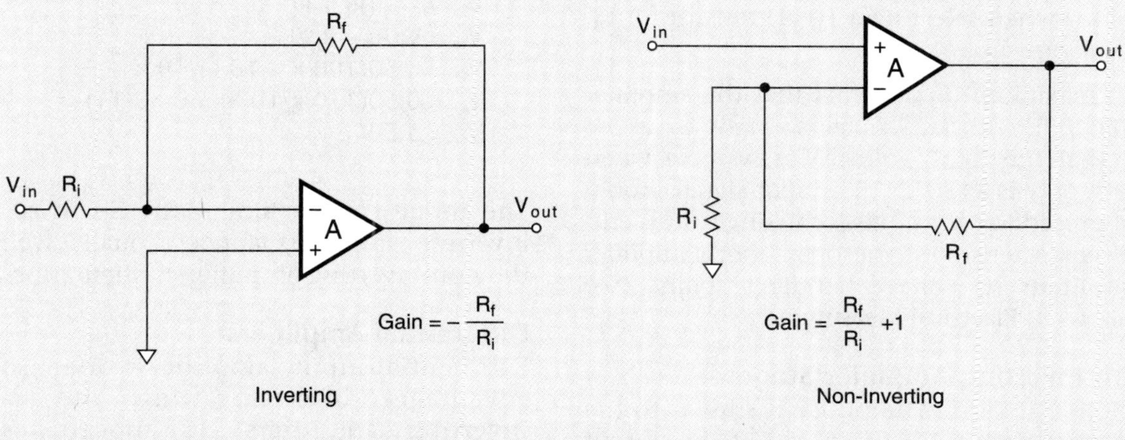

Fig. 5.04. The two basic types of operational amplifiers are called inverting and non-inverting. The stage gain equals the ratio between the feedback and input resistor values.

operational amplifier is in the hundreds of thousands. But the idealized amplifier used to derive the transfer function assumes a gain of infinity to simplify its derivation without introducing significant errors in calculating the stage gain. With such a high stage gain, the input voltage sees only the voltage divider composed of R_f and R_i. The negative sign in the transfer function indicates that the output signal is the inverse polarity of the input. Without deriving the transfer function, the output is calculated from:

EQN 5.01. Inverting Amplifier

$V_o = -V_{in}(R_f/R_i)$
Where:
V_o = output signal, V
V_{in} = input signal, V
R_f = feedback resistor, Ω
R_i = input resistor, Ω

For example, for a 500 mV input signal and a desired output of -5 V:

$V_o = V_{in}(R_f/R_i)$
$V_o/V_{in} = R_f/R_i$
$5/0.50 = R_f/R_i = 10$

Therefore, the ratio between input and feedback resistors should be 10, so R_f must be 100 kΩ when selecting a 10 kΩ resistor for R_i. (See Figure 5.05.)

The maximum input signal that the amplifier can handle without damage is usually about 2 V less than the supply voltage. For example, when the supply is ±15 Vdc, the input signal should not exceed ±13 Vdc. This is the single most critical characteristic of the operational amplifier that limits its voltage handling ability. (See Chapter 4, Electrical Measurements.)

Non-Inverting Amplifier Stages

The non-inverting amplifier is similar to the previous circuit but the phase of the output signal matches the input. Also, the gain equation simply depends on the voltage divider composed of R_f and R_i. (See Figure 5.06.)

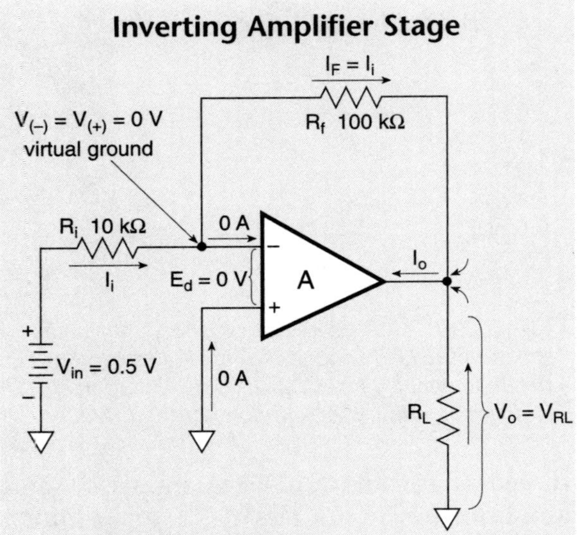

Fig. 5.05. The output polarity of the inverting amplifier is opposite to that of the input voltage. The closed-loop amplification or gain of this stage is A_{cl} = 10, which is the ratio of R_f/R_i or 100 kΩ/10 kΩ.

The simplified transfer function is:

EQN 5.02. Non-Inverting Amplifier

$V_o = V_{in}(R_f + R_i)/R_i$

For the same 500 mV input signal, R_f = 100 kΩ, and R_i = 10 kΩ:

$V_o/V_{in} = (R_f + R_i)/R_i,$
$V_o = V_i(R_f + R_i)/R_i$
$V_o = 0.50(100\ k + 10\ k)/10\ k$
$V_o = 0.50(110\ k/10\ k) = 0.50(11)$
$V_o = 5.5\ V$

The input voltage limitations discussed for inverting amplifiers applies equally well to the non-inverting amplifier configuration.

Differential Amplifiers

Differential-input amplifiers offer some advantages over inverting and non-inverting amplifiers. It appears as a combination of the inverting and non-inverting amplifiers as shown in Figure 5.07. The input signal is impressed between the

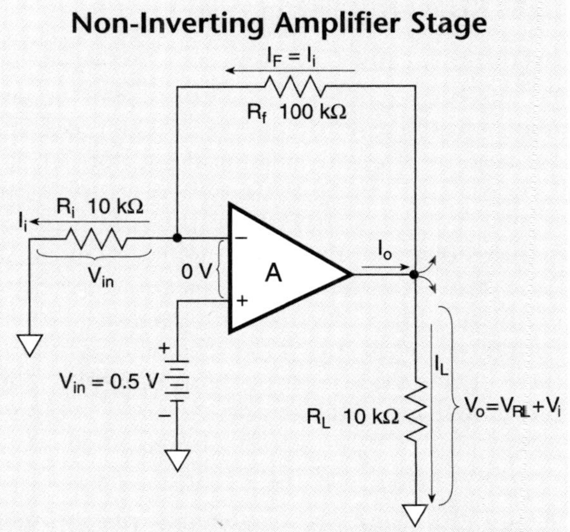

Fig. 5.06. The input and output polarities of the non-inverting amplifier are the same. The gain of the stage is A_{cl} = 11 or $(R_f + R_i)/R_i$.

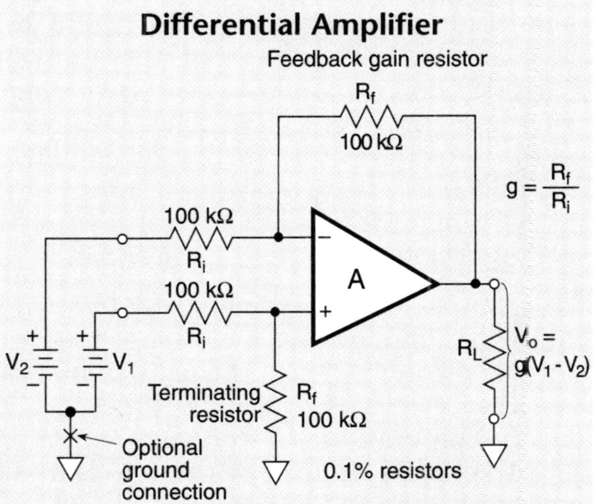

Fig. 5.07. The output voltage of the basic differential amplifier is the difference between the two inputs, or $A_{cl} = g(V_1\text{-}V_2)$, where g is the gain factor. Because all resistors in this example are of equal value, the gain is unity. However, a gain of 10 may be obtained by making the feedback resistor 10 times larger than the input resistors, under the conditions that both feedback resistors are equal and the input resistors are equal.

operational amplifier's positive and negative input terminals and can be isolated from common or a ground pin. The optional ground pin is the key to the amplifier's flexibility. The output signal of the differential input amplifier responds only to the differential voltage that exists between the two input terminals. The transfer function for this amplifier is:

EQN 5.03. Differential Amplifier

$$V_o = (R_f/R_i)(V_1 - V_2)$$

For an input signal of 50 mV, Where: V_1 = 1.050 V and V_2 = 1.000 V

$$V_o = (R_f/R_i)(V_1 - V_2)$$
$$V_o = (100\ \text{k}/100\ \text{k})(0.05\ \text{V})$$
$$V_o = 0.05\ \text{V}$$

For a gain of 10 where R_f = 100 k and R_i = 10 k:

$$V_o = (R_f/R_i)(V_1 - V_2)$$
$$V_o = (100\ \text{k}/10\ \text{k})(0.05\ \text{V})$$
$$V_o = 0.50\ \text{V}$$

The major benefit of the differential amplifier is its ability to reject any voltages that are common to both inputs while amplifying the difference voltage. The voltages that are common to both inputs are appropriately called common mode voltages (V_{cm} or CMV). The common-mode voltage rejection quality can be demonstrated by connecting the two inputs together and to a voltage source referenced to ground. Although a voltage is present at both inputs, the differential amplifier responds only to the difference, which in this case is zero. The ideal operational amplifier then, yields zero output volts under this arrangement. (See the following section on Instrumentation Amplifiers, High-Common-Mode Amplifiers.)

Programmable-Gain Amplifiers

Programmable gain amplifiers are typically non-inverting operational amplifiers with a digitally controlled analog switch connected to several resistors in its feedback loop. An external computer or another logic or binary signal controls the addressable inputs of the analog switch so it selects a certain resistor for particular gain. (See Figure 5.08.) The data acquisition system's signal conditioners sense the input signal amplitude and automatically send the proper binary code to the programmable gain amplifier (PGA) to increase the gain for a low signal, or decrease the gain for a large signal. The input signal then can be measured and displayed without distortion.

Programmable-Gain Amplifier

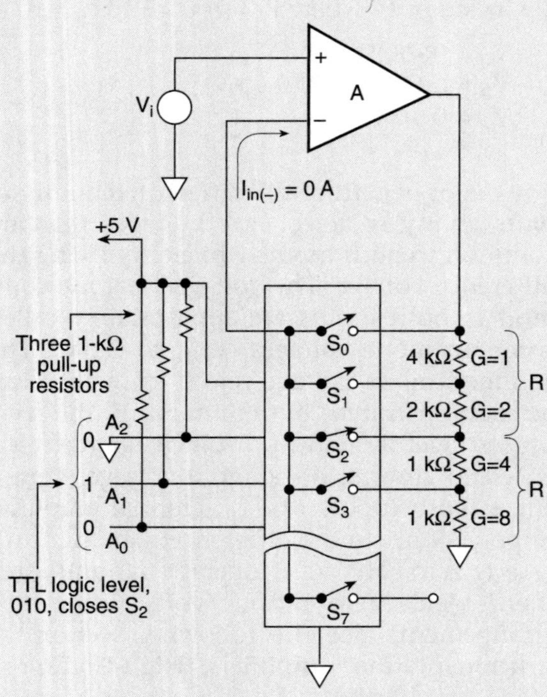

Fig. 5.08. The non-inverting amplifier is configured for programmable gain and controlled by the binary input signals from a computer to the addressable inputs of the analog switch.

INSTRUMENTATION AMPLIFIERS

A Fundamental Problem

Because signal levels from some transducers may be just a few microvolts, special problems concerning ground loops and spurious interference frequently arise when amplifying them. Other transducers provide output signals from differential signal sources to minimize grounding problems and reduce the effect of common-mode interfering signals. Amplifiers used in these applications must have extremely low input current, drift, and offset voltage; stable and accurate voltage gain; and high input impedance and common-mode rejection.

Although common integrated operational amplifiers with several stages and extremely tight resistor ratios are often used, specially designed instrumentation amplifiers are preferred for these applications. The high-performance operational amplifiers still use basic circuits but ensure that they provide extremely high common-mode rejection and don't need high precision matched resistors to set the gain. Many instrumentation amplifiers are designed for special applications and provide unique features to increase their accuracy and stability for those applications.

For example, the functional block following the switching network in a data acquisition system, Figure 5.09, is an instrumentation amplifier with several critical functions. It rejects common-mode voltages, amplifies signal voltages, minimizes the effect of the multiplexer on resistance, and drives the ADC input.

High Common-Mode Amplifiers

The common mode voltage is defined as the voltage applied from analog common to both inputs when the inputs are identical. (See Figure 5.10.) However, when the two input voltages are different, 4.10 and 4.20 V for

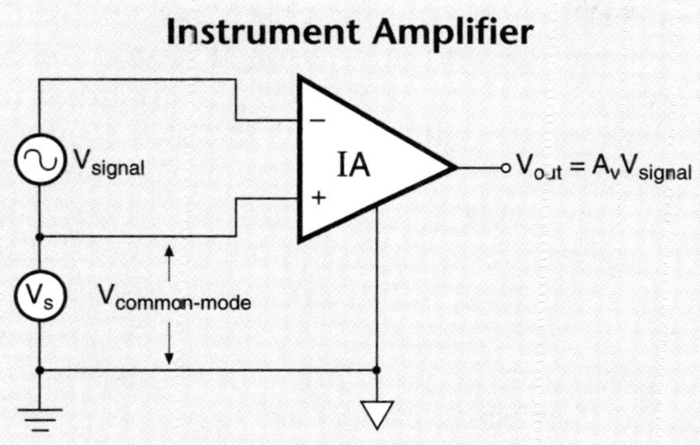

Fig. 5.09. An instrumentation amplifier is typically a differential input operational amplifier with a high common-mode rejection ratio.

example, the common mode voltage, V_{cm}, is 4.10 V and the differential voltage between the two is 0.10 V. Ideally, the instrumentation amplifier ignores the common mode voltage and amplifies only the difference between the

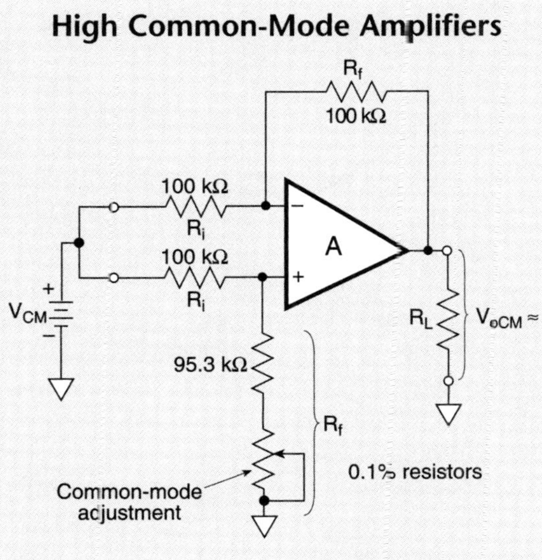

Fig. 5.10. The common mode voltage rejection is measured with the two inputs shorted together and a voltage applied to the node. The potentiometer is then adjusted for minimum (V_{OCM}) output from the amplifier, signifying that the best balance was reached between the two inputs.

two inputs. The degree to which the amplifier rejects common mode voltages is given by a parameter called common mode rejection ratio CMRR. (See Chapter 4, When to Make Differential Measurements.) The ability of an instrumentation amplifier (IA) to reject high common mode voltages is sometimes confused with its ability to reject high voltages. (See Chapter 4, Electrical Measurements).

The signal voltages measured are frequently much smaller than the maximum allowed input of the system's ADC. For example, a 0 to 100 mV signal is much smaller than the 0 to 5 V range of a typical ADC. A gain of 50 is needed to obtain the maximum practical resolution for this measurement. Instrumentation amplifiers are capable of gains from 1 to more than 10,000, but in multiplexed systems, the gains are usually restricted to a range of 1 to 1,000.

Measurement errors come from the non-ideal ON resistance of analog switches added to the impedance of any signal source. But the extremely high input impedance of the IA minimizes this effect. The input stage of an IA consists of two voltage followers, which have the highest input impedance of any common amplifier configuration. The high impedance and extremely low bias current drawn from the input signal generate a minimal voltage drop across the analog switch sections and produce a more accurate signal for the IA input.

The instrumentation amplifier has low output impedance, which is ideal for driving the ADC input. The typical ADC does not have high or constant input impedance, so the preceding stage must provide a signal with the lowest impedance practical.

Some instrumentation amplifiers have limitations concerning offset voltage, gain error, limited bandwidth, and settling time. The offset voltage and gain error can be calibrated out as part of the measurement, but the bandwidth and settling time are parameters that limit the frequencies of amplified signals and the frequency at which the input switching system can switch channels between signals. A series of steady dc voltages applied to an instrumentation amplifier in rapid succession generates a difficult composite signal to amplify. The settling time of the amplifier is the time necessary for the output to reach final amplitude to within some small error (often 0.01%) after the signal is applied to the input. In a system that scans inputs at 100 kHz, the total time spent reading each channel is 10 µs. If analog-to-digital conversion requires 8 µs, settling time of the input signal to the required accuracy must be less than 2 µs.

Although calibrating a system can minimize offset voltage and gain error, it is not always necessary to do so. For example, an amplifier with an offset voltage of 0.5 mV and a gain of 2 measuring a 2 V signal develops an error of only 1 mV in 4 V on the output, or 0.025%. By comparison, an offset of 0.5 mV and a gain of 50 measuring a 100 mV signal develop an error of 25 mV in 5 V or 0.5%. Gain error is similar. A stage gain error of 0.25% has a greater overall effect as gain increases producing larger errors at higher gains and minimal errors at unity gain. System software can generally handle known calibration constants with mx+b routines, but some

Integrated Instrumentation Amplifiers

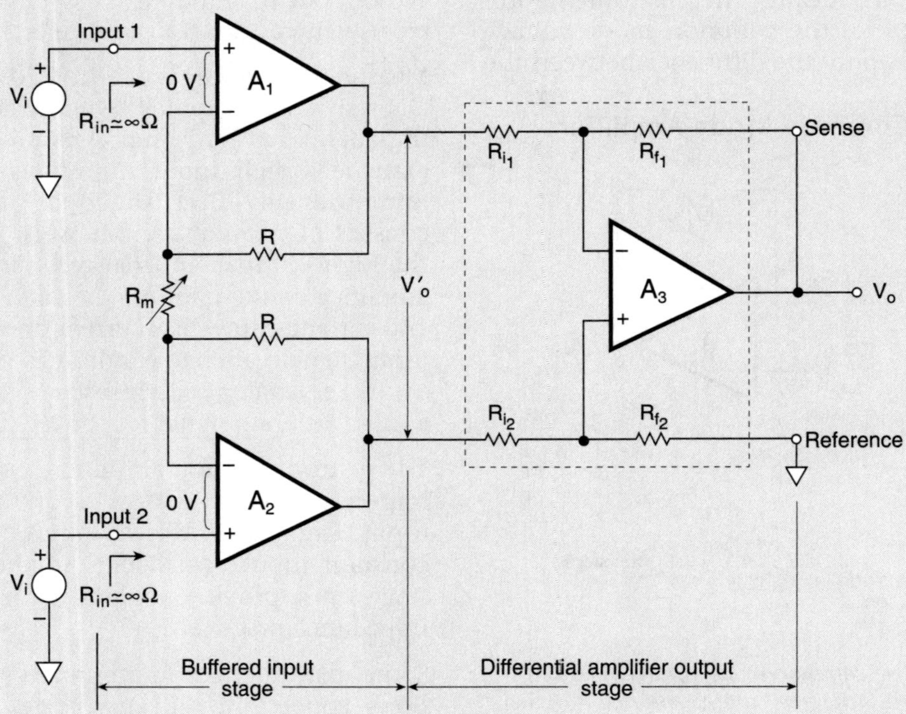

Fig. 5.11. The instrumentation amplifier exhibits extremely high impedance to the inputs V_1 and V_2. Resistor R_m adjusts the gain, and the single-ended output is a function of the difference between V_1 and V_2.

measurements are not critical enough to justify the effort.

Integrated Instrumentation Amplifiers

Integrated instrumentation amplifiers are high-quality op amps that contain internal precision feedback networks. They are ideal for measuring low-level signals in noisy environments without error, and amplifying small signals in the midst of high common-mode voltages. Integrated instrumentation amplifiers are well suited for direct connection to a wide variety of sensors such as strain gages, thermocouples, RTDs, current shunts, and load cells. They are commonly configured with three op amps – two differential inputs and one differential output amplifier. (See Figure 5.11.) Some have built-in gain settings of 1 to 100 and others are programmable.

Programmable-Gain Instrumentation Amplifiers

A special class of instrumentation amplifiers, called programmable-gain instrumentation amplifiers (PGIA), switch between fixed gain levels at high speeds for different input signals delivered by the input switching system. The same digital control circuitry that selects the input channel also can select a gain range. The principle of operation is the same as that described above for Programmable Gain Amplifiers.

FILTERING

The three most common types of filters are called Butterworth, Chebyshev, and Bessel. (See Figures 5.12A, B, and C.) Each type has unique characteristics that make it more suitable for one application than another. All may be used for high-pass, low-pass, band-pass, and band-reject applications, but they have different response profiles. They may be used in passive or active filter networks.

The Butterworth filter has a fairly flat response in the pass-band for which it is

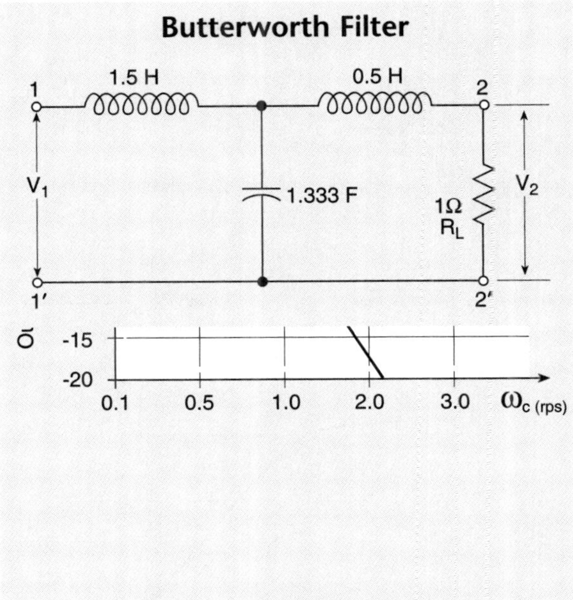

Fig. 5.12A. Butterworth filter and response characteristics

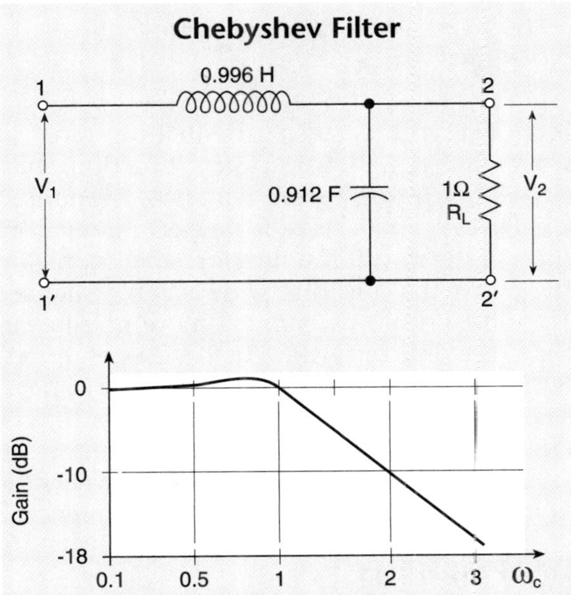

Fig. 5.12B. Chebyshev filter and response characteristics

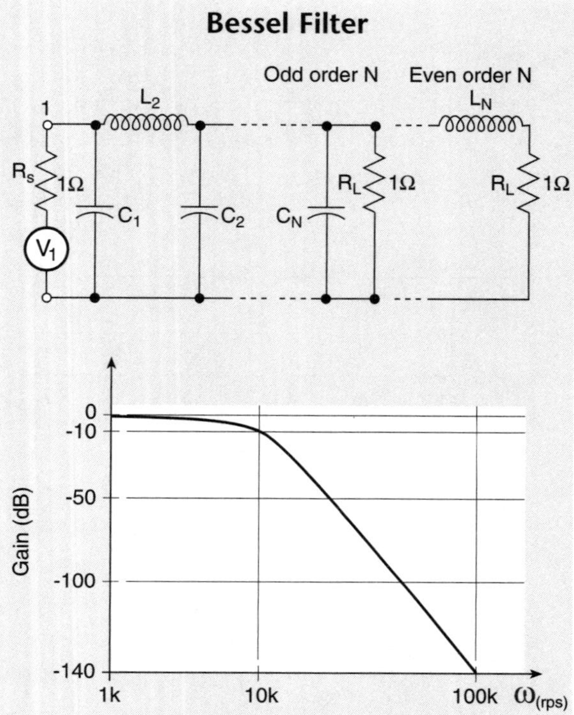

Fig. 5.12C. Bessel filter and response characteristics

intended and a steep attenuation rate. It works quite well for a step function, but shows a non-linear phase response. Chebyshev filters have a steeper attenuation than Butterworth, but develop some ripple in the pass band and ring with a step response. The phase response is much more non-linear than the Butterworth. Finally, Bessel filters have the best step response and phase linearity. But to be most useful, Bessel filters need to have a high order (number of sections) to compensate for their slower rate of attenuation beyond the cut-off frequency.

Low-Pass Filters

Low pass filters attenuate higher frequencies in varying degrees depending on the number of stages and the magnitude of the high frequency relative to the corner frequency. An amplifier stage doesn't need high bandwidth when the measured signal is at a much lower frequency. In fact, the design is intended to eliminate excessive bandwidth in all circuits, which reduces noise. One major benefit of individual signal conditioning stages for low-level sensors (as opposed to multiplexed stages) is to include low-pass filtering on a per-channel basis in the signal path. In a multiplexed circuit (an amplifier which is being shared by multiple low-level dc signals), the main signal path generally cannot operate as a low-pass filter due to the fast settling time necessary in multi-plexed systems. (See Figure 5.13A.)

The best place for low-pass filters is in the individual signal path before buffering and multiplexing. (See Figure 5.13B.) For small signals, amplifying with an instrumentation amplifier prior to filtering lets an active low-pass filter operate at optimum signal-to-noise ratios. Figure 5.13C illustrates a typical amplifier-filter-multiplexer configuration.

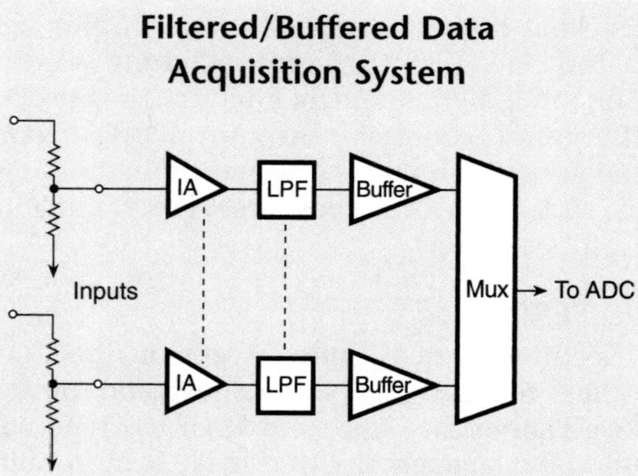

Fig.5.13A. An instrumentation amplifier, a low-pass filter, and a unity gain buffer amplifier placed between the sensor output and MUX input transforms the sensor's high output impedance to a low impedance. This is required by the MUX to maintain accuracy.

Simple RC Filter

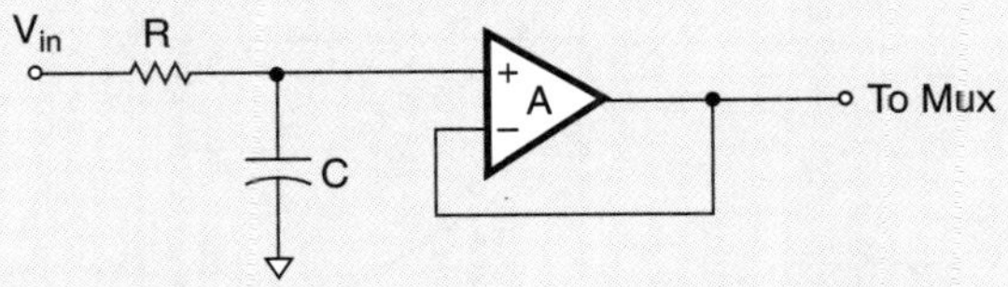

Fig.5.13B. Low-pass filters inserted in each channel as needed simultaneously reduce the bandwidth and noise while passing the targeted, lower frequency signals.

High-Pass Filters

High-pass filters operate in reverse to low-pass filters. They attenuate the lower frequencies and are needed when low-frequency interference can mask high-frequency signals carrying the desired information or data. Low frequency electrical interference sometimes couples into the system from 50 or 60-Hz power lines. Similarly, when analyzing a machine for vibration, the desired signals can be corrupted by low frequency mechanical interference from the vibrating laminations of a power transformer mounted to its frame. Moreover, a combination of high-pass and low-pass filters may be used to create a notch filter to attenuate a narrow band of frequencies, such as 50 to 60 Hz and their first harmonic. A 3-pole high-pass filter is shown in Figure 5.14.

High-Pass Filter

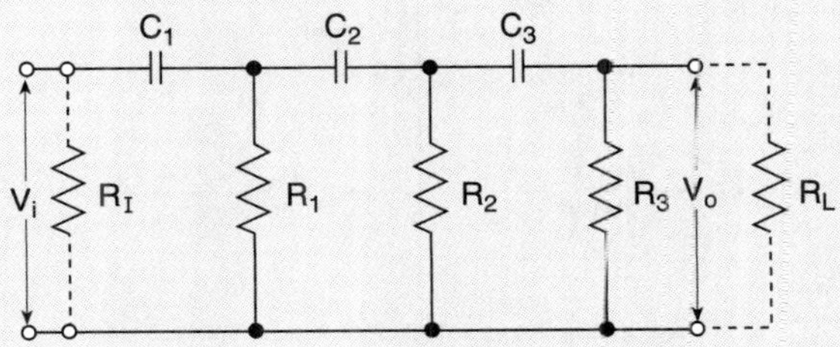

Fig. 5.14. The high-pass filter is designed to have a lower corner frequency near zero and the cutoff frequency at some higher value. The number of capacitor/resistor pairs determines the number of poles and the degree of cutoff sharpness.

Amplifier/Filter/MUX Combination

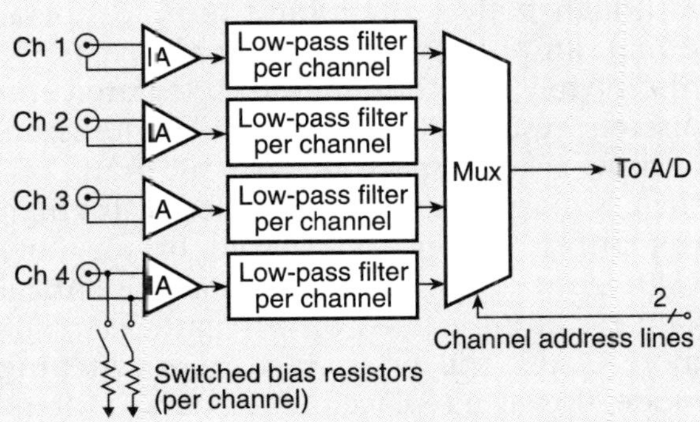

Fig. 5.13C. A typical production type data acquisition system contains an instrumentation amplifier and a low-pass filter in each channel preceding the MUX. The amplifier boosts low-level signals sufficiently well to let the active filter work at an optimum signal-to-noise ratio.

Passive vs. Active Filters

Passive filters comprise discrete capacitors, inductors, and resistors. As the frequencies propagate through these networks, two problems arise: the desired signal is attenuated by a relatively small amount, and when connected to a load, the original filtering characteristics change. However, active filters circumvent these problems. (See Figure 5.15.) They comprise operational amplifiers built with both discrete and integrated resistors, capacitors, and inductors. They can provide the proper pass-band (or stop-band) capability without loading the circuit, attenuating the desired signals, or changing the original filtering characteristics. The advantages of such a filter is; it is easier to design, and it has finer tuning, fewer parts count, and improved filter characteristics.

Active Filters

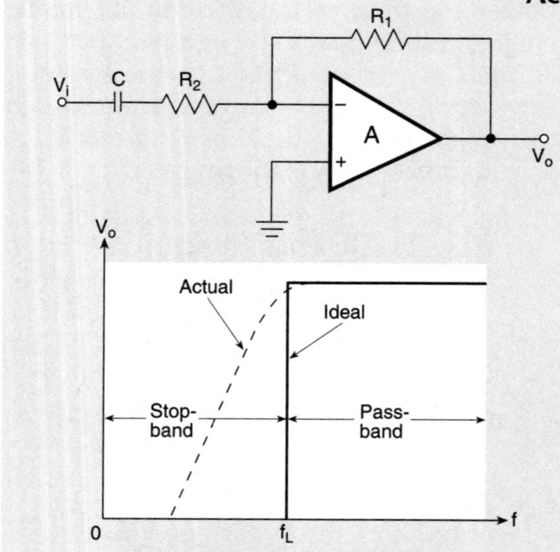

High-Pass Filter

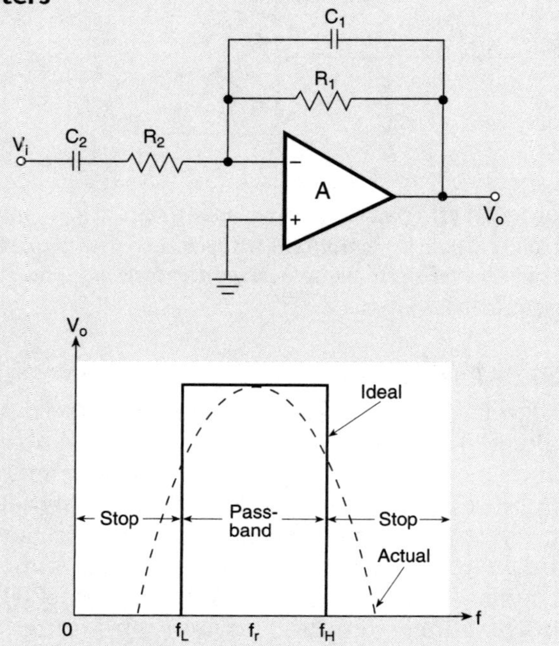

Band-Pass Filter

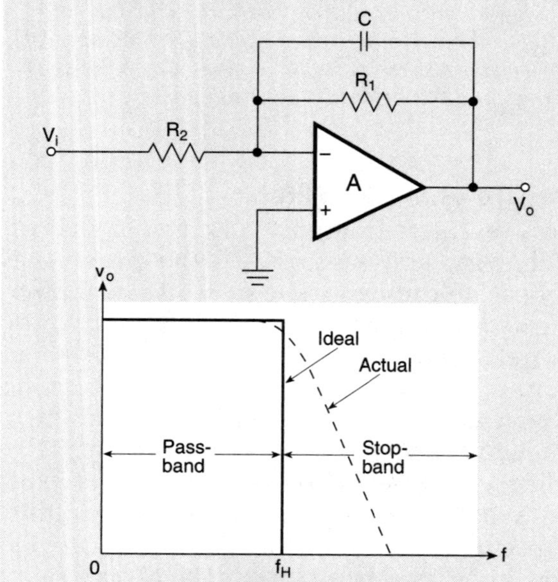

Low-Pass Filter

Fig. 5.15. *Passive filters tend to change in frequency-cutoff characteristics with a change in load. To prevent this, an active device such as a transistor or op amp isolates the last pole from the load to maintain stable filter characteristics.*

Switched-Capacitance Filters

Although active filters built around operational amplifiers are superior to passive filters, they still contain both integrated and discrete resistors. Integrated-circuit resistors occupy a large space on the substrate and their values can't be made easily to high tolerances, either in relative or absolute values. But capacitors with virtually identical values can be formed on integrated circuits more easily, and when used in a switching mode, they can replace the resistors in filters.

The switched-capacitor filter is a relatively recent improvement over the traditional active filter. James Clerk Maxwell compared a switched capacitor to a resistor in a treatise in 1892, but only recently has the idea taken hold in a zero-offset electronic switch and a high input impedance amplifier. The switched-capacitor concept is now used in extremely complex and accurate analog filter circuits.

The theory of operation is depicted in Figure 5.16. It shows that with S_2 closed and S_1 open, a charge from V_2 accumulates on C. Then, when S_2 opens, S_1 closes, and the capacitor transfers the charge to V_1. This process repeats at a particular frequency, and the charge becomes a current by definition, that is, current equals the transfer of charge per unit time.

Switched-Capacitor Filters

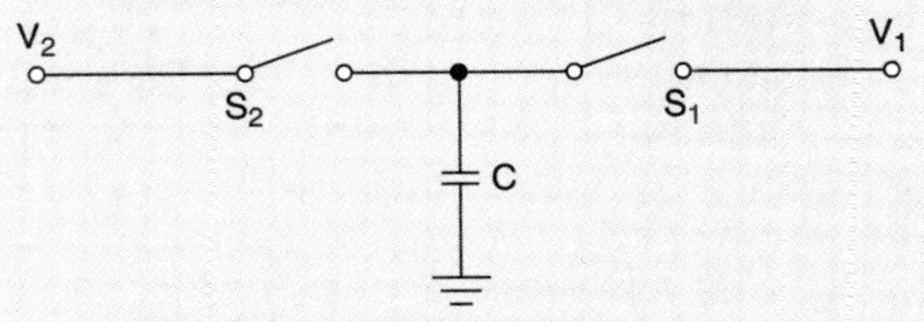

Fig. 5.16. Because resistors have wider tolerances and require more substrate area than capacitors, a technique that uses multiple precision capacitors to replace resistors in filters is called a switched capacitor circuit.

The derivation of the equation is beyond the scope of this handbook, but it can be shown that the equivalent resistor may be determined by:

EQN 5.04 Switched-Capacitance Filters

$$(V_2 - V_1)/i = 1/(fC) = R$$

Where:
V_2 = voltage source 2, V
V_1 = voltage source 1, V
i = equivalent current, A
f = clock frequency, Hz
C = capacitor, F
R = equivalent resistor, Ω

The above equation states that the switched capacitor is identical to a resistor within the constraints of the clock frequency and fixed capacitors. Moreover, the equivalent resistor's effective value is inversely proportional to the frequency or the size of the capacitor.

ATTENUATION

Voltage Dividers

Most data acquisition system inputs can measure voltages only within a range of 5 to 10 V. Voltages higher than this must be attenuated. Straightforward resistive dividers can easily attenuate any range of voltages (see Figure 5.17), but two drawbacks complicate this simple solution. First, voltage dividers present substantially lower impedances to the source than direct analog inputs. Second, their output impedance is much too high for multiplexer inputs. For example, consider a 10:1 divider reading 50 V. If a 900 kΩ and a 100 kΩ resistor are chosen to provide a 1 MΩ load to the source, the impedance seen by the analog multiplexer input is about 90 kΩ, still too high for the multiplexed reading to be accurate. When the values are both downsized by a factor of 100 so the output impedance is less than 1 kΩ, the input impedance seen by the measured source is 10 kΩ, or

Attenuator/Buffer

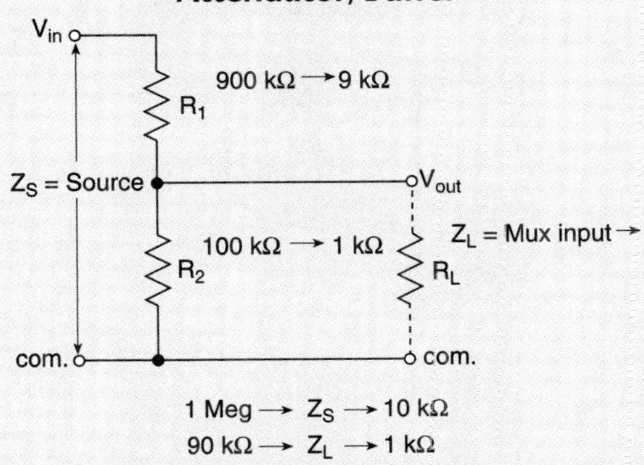

Fig. 5.17. When the input signal exceeds about 10 V, the divider drops the excess voltage to prevent input amplifier damage or saturation.

2 kΩ/V, which most instruments cannot tolerate in a voltage measurement. Hence, simple attenuation is not practical with multiplexed inputs.

Buffered Voltage Dividers

The low impedance loading effect of simple voltage dividers can be overcome using unity-gain buffer amplifiers on divider outputs. A dedicated unity-gain buffer has high-input impedance in the MΩ range and does not load down the source, as does the network in the previous example. Moreover, the buffers' output impedance is extremely low, which is necessary for the multiplexed analog input. (See Figure 5.18.)

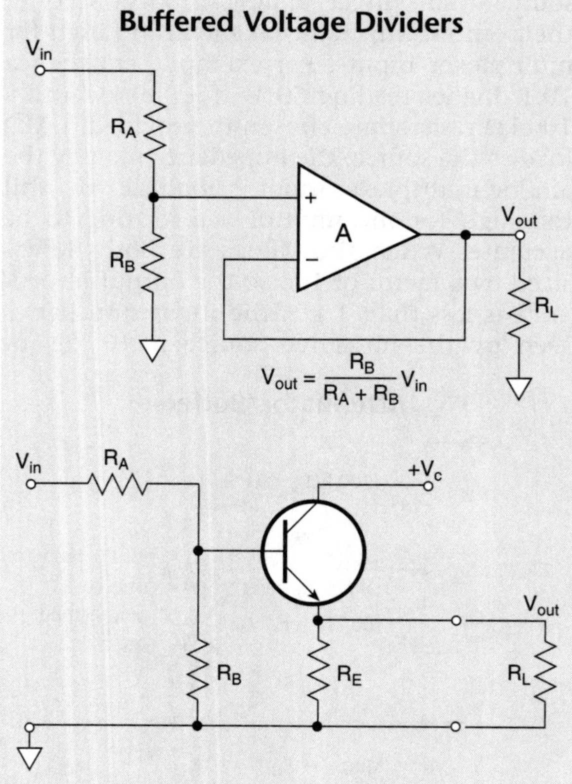

Fig. 5.18. An op amp or a transistor serves as an impedance matching buffer to prevent the load from affecting the divider's output voltage.

Balanced Differential Dividers

Not all voltage divider networks connect to a ground or a common reference point at one end. Sometimes, a balanced differential divider is a better solution for driving the data acquisition system's input terminals. (See Figure 5.19.) In this case, the CMRR of the differential amplifier effectively reduces the common-mode noise that can develop between different grounds in the system.

High-Voltage Dividers

Some data acquisition systems employ special input modules containing high-voltage dividers that can easily measure up to 1,200 V. They are properly insulated to handle the high voltage and have resistor networks to select a number of different divider ratios. They also contain internal trim potentiometers to calibrate the setup to extremely close tolerances.

Compensated Voltage Dividers and Probes

Voltage divider ratios applied to dc voltages are consistently accurate over relatively long distances between the divider network and data acquisition system input when the measurement technique eliminates the dc resistance of the wiring and cables. These techniques include a second set of input measuring leads separate from those that apply power to the divider such as the Kelvin connection discussed in Chapter 4.

Voltage dividers used on ac voltages, however, must always compensate for the effective capacitance between the conductors and ground or common, even when the frequency is as low as 60 Hz. It's not uncommon to find that when the ac voltages are calibrated to within 0.01% at the divider network, the voltages reaching the data acquisition system input terminals may be out of tolerance by as

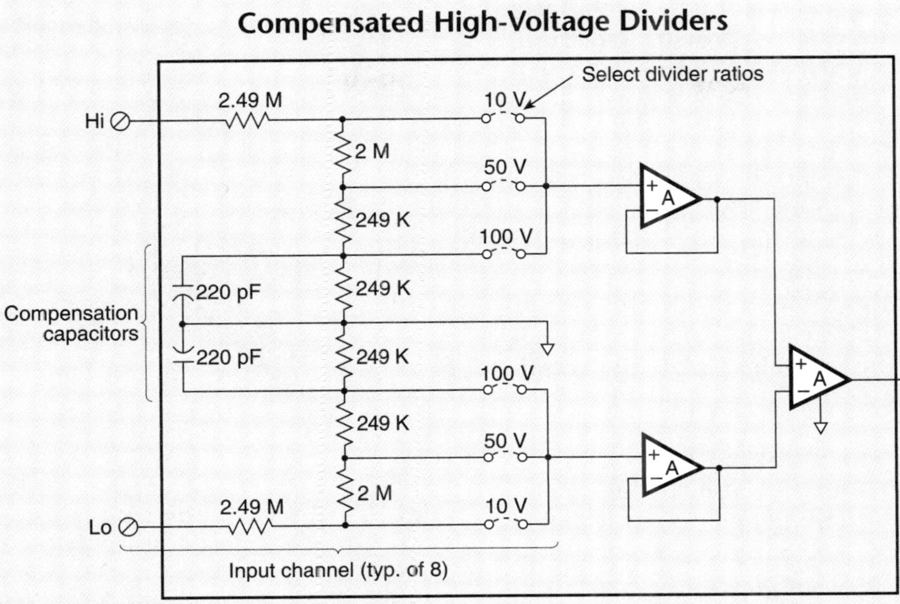

Fig. 5.19. A typical high-voltage input card for a data acquisition system contains a balanced differential input and jumpers for selecting an input voltage range of 10 V, 50 V, or 100 V. The input circuit also compensates for long lead wire capacitance that tends to form an ac voltage divider, which would reduce accuracy.

much as 5%. This is because the lead capacitance enters into the divider equation. One way around the problem is to shunt the data acquisition input terminals (or the divider network) with a compensating capacitor. For example, oscilloscope probes contain a variable capacitor precisely for this purpose. The probe capacitor is adjusted to match the oscilloscope's input impedance and as a result, it faithfully passes the leading edge of the oscilloscope's built-in 1,000-Hz square-wave generator without undershoot or overshoot.

ISOLATION

When Isolation Is Required

Frequently, data acquisition system inputs must measure low-level signals where relatively high voltages are common, such as in motor controllers, transformers, and motor windings. In these cases, isolation amplifiers can measure low-level signals among high common-mode voltages, break ground loops, and eliminate source ground connections without subjecting operators and equipment to the high voltage. They also provide a safe interface in a hospital between a patient and a monitor or between the source and other electronic instruments and equipment. Yet other applications include precision bridge isolation amplifiers, photodiode amplifiers, multiple-port thermocouple and summing amplifiers, and isolated 4 to 20 mA current-control loops.

Isolation Amplifiers

Isolation amplifiers are divided into input and output sections, galvanically isolated from each other. Several techniques provide the isolation; the most widely used include capacitive, inductive, and optical means. The isolation voltage rating is usually 1,200 to 1,500 V_{ac}, at 60 Hz with a typical input signal range of ±10 V. They normally have a high isolation mode rejection (IMR) ratio of around

140 dB. Because the primary job of relatively low-cost amplifiers is to provide isolation, many come with unity gain. But more expensive units are available with adjustable or programmable gains. (See Figure 5.20.)

One benefit of an isolation amplifier is that it eliminates ground loops. The input section's signal-return, or common connection is isolated from the output signal ground connection. Also, two different power supplies are used, V_{CC1} and V_{CC2}, one for each section, which further helps isolate the amplifiers. (See Figure 5.21.)

Analog Isolation Modules

Analog isolation amplifiers use all three types of isolation between input and output sections: capacitive, optical, and magnetic. One type of capacitively coupled amplifier modulates the input signal and couples it across a capacitive barrier with a value determined by the duty

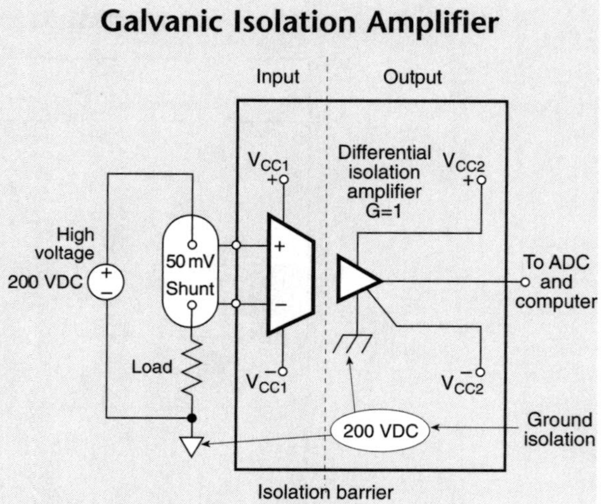

Fig. 5.21. Galvanic isolation can use any one of several techniques to isolate the input from the output circuitry. The goal is to allow the device to withstand a large common mode voltage (CMV) between the input and output signal and power grounds.

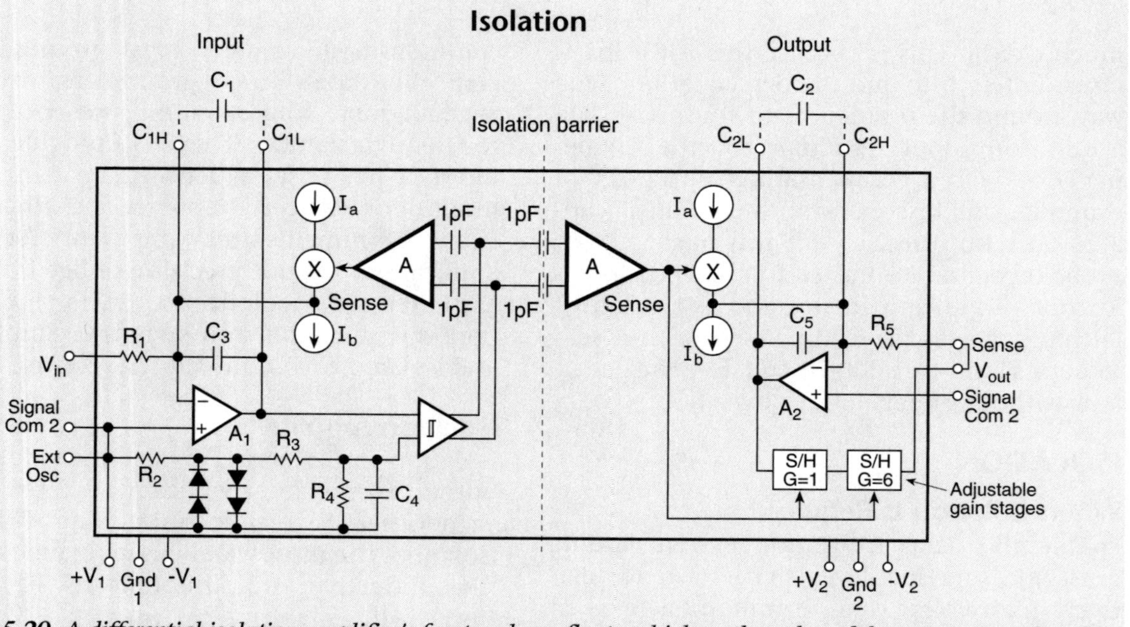

Fig. 5.20. A differential isolation amplifier's front end can float as high as the value of the common mode voltage (CMV) rating without damage or diminished accuracy. The isolation barrier in some signal conditioners can withstand from 1,500 to 2,200 Vdc.

cycle. (See Figure 5.22.) The output section demodulates the signal, restores it to the original analog input equivalent, and filters the ripple component, which resulted from the demodulation process. After the input and output sections of the integrated circuit are fabricated, a laser trims both stages to precisely match their performance characteristics. Then they are mounted on either end of the package separated by the isolation capacitors. Although the schematic diagram of the isolation amplifier looks quite simple, it can contain up to 250 or more integrated transistors.

Another isolation amplifier optically couples the input section to the output section through an LED transmitter and receiver pair as shown in Figure 5.23. An ADC converts the input signal to a time-averaged bit stream and transmits it to the output section via the LED. The output section converts the digital signal back to an analog voltage and filters it to remove the ripple voltage.

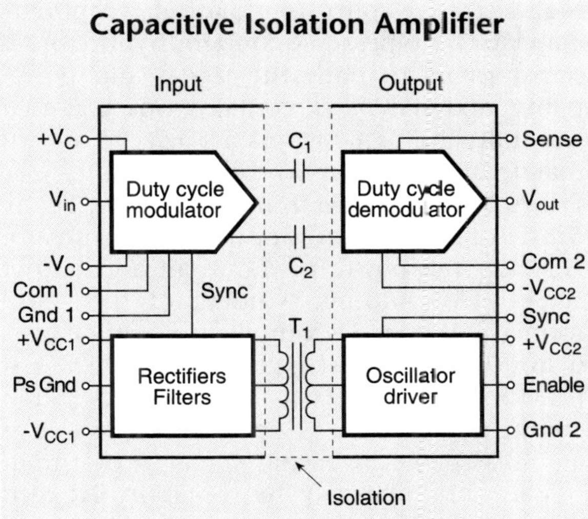

Fig. 5.22. The isolation barrier in this amplifier protects both the signal path and the power supply from CMV breakdown. The signal couples through a capacitor and the power through an isolation transformer.

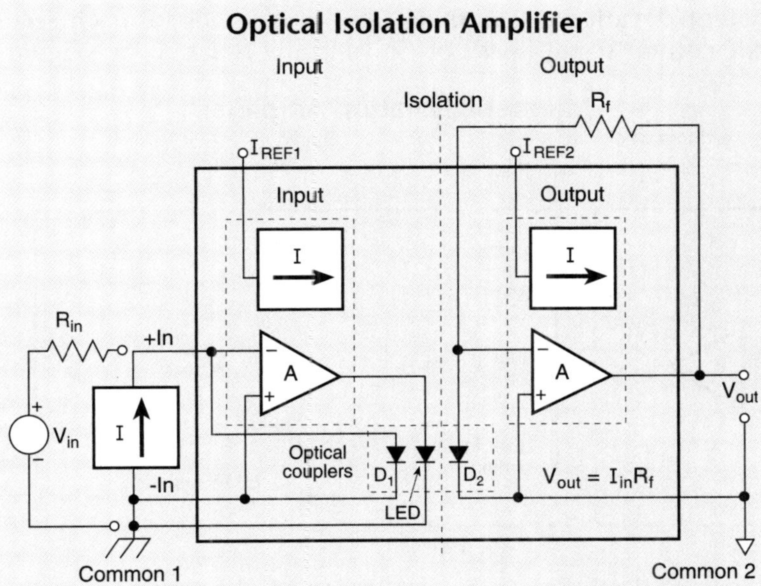

Fig. 5.23. This is a simplified diagram of a unity gain current amplifier using optical couplers between input and output stages to achieve isolation. The output current passing through the feedback resistor (Rf) generates the output voltage.

Magnetically coupled isolation amplifiers come in two types. One contains hybrid toroid transformers in both the signal and power paths, and the other contains one coil that transmits the signal across a barrier to a giant magnetoresistor (GMR) bridge circuit. (See Figure 5.24A.) In the transformer type, Figure 5.24B, the rectified output of a pulse generator (T1) supplies power to the input and output stages (T3). Another winding of the transformer (T2) operates a modulator and demodulator that carry the signal across the barrier. It provides from 1,000 to 3,500 Vdc isolation among the amplifier's three grounds, as well as an isolated output signal equal to the input signal with total galvanic isolation between input and output terminals.

The second type, the giant magnetoresistor amplifier uses the same basic technology as does high-speed hard disc drives. The coil generates a magnetic field with strength proportional to its input drive current signal, and the dielectric GMR amplifies and conditions it. Ground potential variations at the input do not generate current so they are not detected by the magnetoresistor structure. As a result, the output signal equals the input signal with complete galvanic isolation. These units are relatively inexpensive, and can withstand from 1,000 to 3,500 V_{dc}. Full-power signal frequency response is less than 2 kHz, but small signal response is as much as 30 kHz.

Digital Method of Isolation

Digital isolation packages are similar in many respects to analog amplifiers. They transmit digital data across the isolation barrier at rates up to 80 Mbaud, and some can be programmed to transmit data in either direction, that is, through input to output or output to input terminals. Data, in the form of complementary pulses, couple across the barrier through high-voltage capacitors or air-core inductors. Faraday shields usually surround the inductors or capacitors to prevent false triggering from external fields. The receiver restores the pulses to the original standard logic levels. As with analog amplifiers, the power supplies for each section are also galvanically isolated. (See Figure 5.25.)

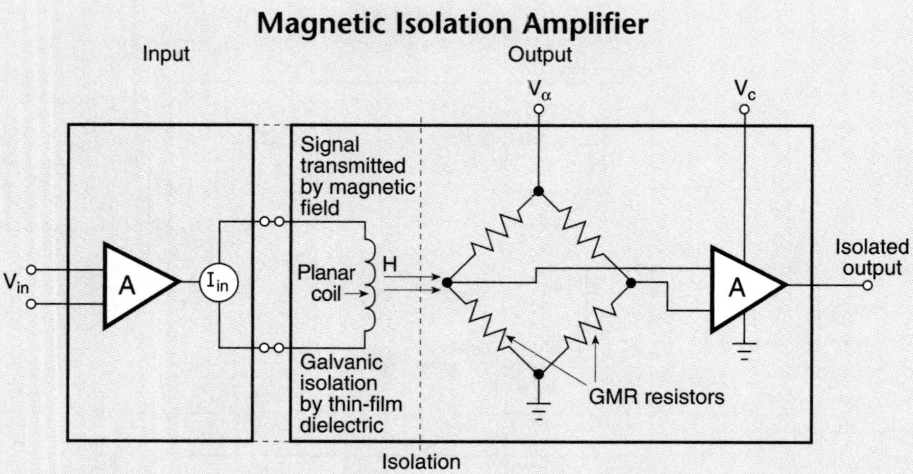

Fig. 5.24A. Magnetic couplers transfer signals through a magnetic field across a thin film dielectric. In this case, a magneto-resistance bridge circuit sensitive to the field exhibits a large change in resistance when exposed to the magnetic field from a small coil sitting above it.

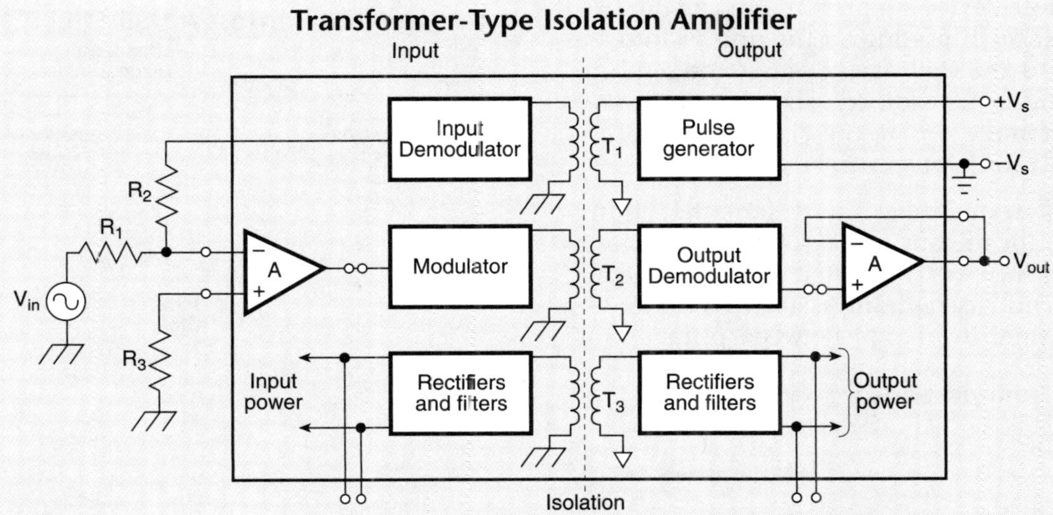

Fig. 5.24B. The transformer-coupled isolation amplifier uses separate power supplies for the input and output stages, which are isolated by virtue of the individual transformer windings. This provides the input and output stages with decoupled ground or common returns. In addition, the modulator/demodulator transfers the measured signal across the barrier via other transformer windings for complete galvanic isolation between input and output.

Inherently Isolated Sensors

In addition to directly measuring voltage, current, and resistance, which require some degree of isolation, certain sensors that measure other quantities are inherently isolated by virtue of their construction or principle of operation. The most widely used sensors measure position, velocity, pressure, temperature, acceleration, and proximity. They also use a number of different devices to measure these quantities, including potentiometers, LVDTs (linear variable differential transformers), optical devices, Hall-effect devices, magnetic devices, and semiconductors.

Hall-effect devices, for example, measure magnetic fields, and are electrically insulated from the magnetic source that they are designed to measure. The insulation may be air or another material such as plastic or ceramic, and the arrangement essentially isolates them from ground loops and high voltages. Figures 5.26A and B illustrate two applications where Hall-effect devices measure speed. The first senses the alternating magnetic field directly from the revolving wheel. In the second application, a permanent magnet sitting behind the

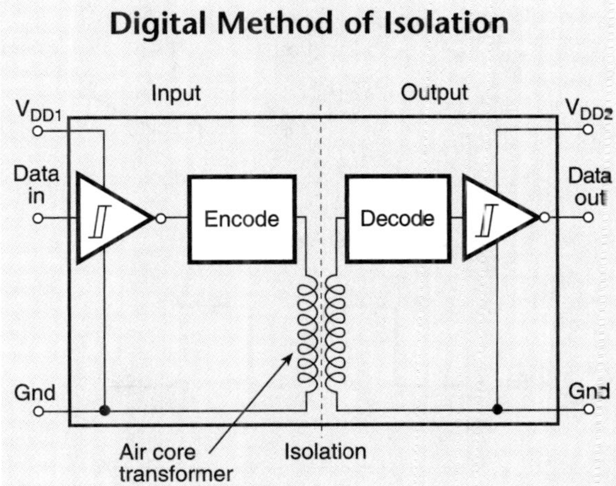

Fig. 5.25. Yet another isolation method specifically intended for digital circuits employs a high-speed CMOS encoder and decoder at the input and output, coupled with a monolithic air-core transformer.

Hall effect device supplies the magnetic field. The gear teeth passing by the unit disturb the field, and the Hall device senses the resulting fluctuations. In addition, Figure 4.08 shows a third example, where the Hall-effect device is used in a wattmeter circuit.

Current transformers and potential transformers for measuring ac voltage and current are also inherently isolated between primary and secondary windings. (See Figure 5.27.) Transformer insulation between primary and secondary can be made to withstand thousands of volts and have extremely low leakage values. The turns ratio also is easy to select for stepping down a high voltage to a lower standard voltage of 5 to 10-Vac.

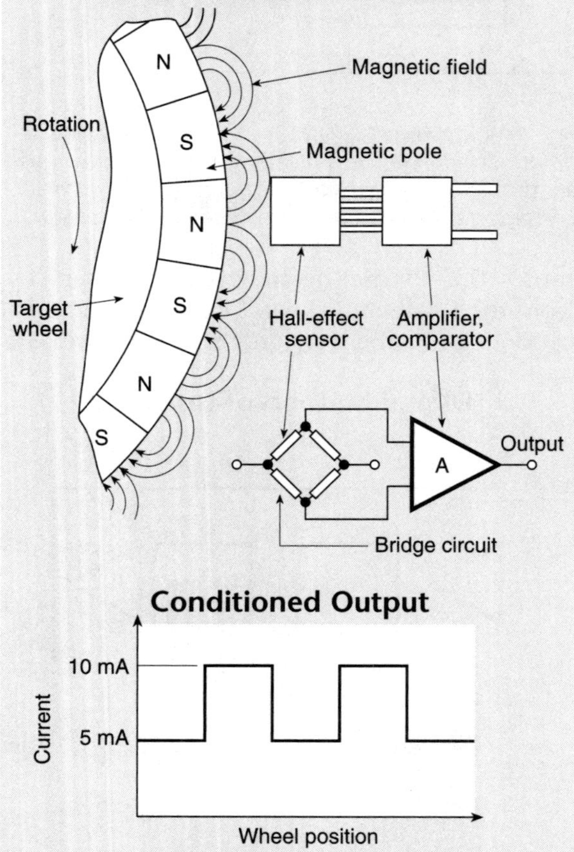

Fig. 5.26A. This Hall-Effect sensor is switched with a series of alternating magnets in a target wheel. Each pass between N and S magnets changes the sensor's state.

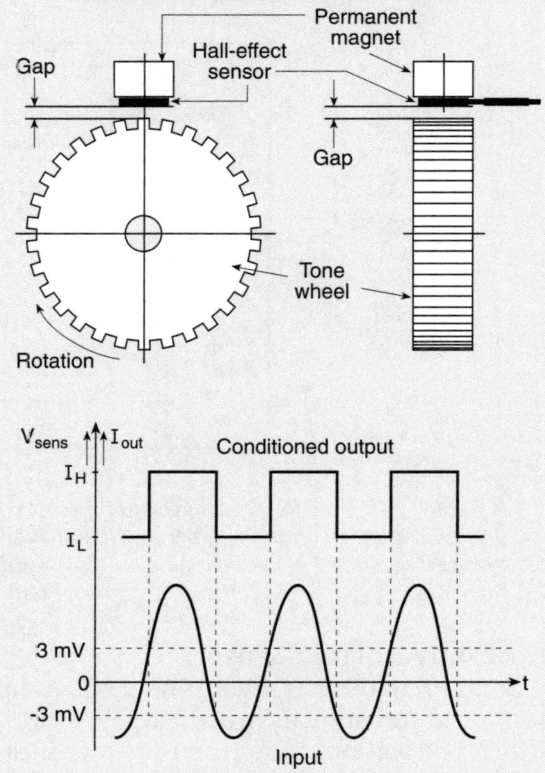

Fig. 5.26B. This Hall-Effect sensor configuration uses a bias magnet and a tone wheel that modulates the magnetic field intensity to produce an output signal.

Other sensors include magnetic pickups composed of wire coils wound around a permanent magnetic core. A ferrous metal passing over one end of the coil disturbs the magnetic flux and generates a voltage at the coil terminals. The sensor does not require a separate power supply, and the output voltage is typically small enough to require only ordinary signal conditioners. (See Figure 5.28.)

Piezoelectric materials and strain gages are typically used for measuring acceleration, and are inherently isolated from the objects on which they are mounted by virtue of their protective housings. High-voltage insulation and magnetic shielding may be added to the

Isolated Sensor – Current Transformer

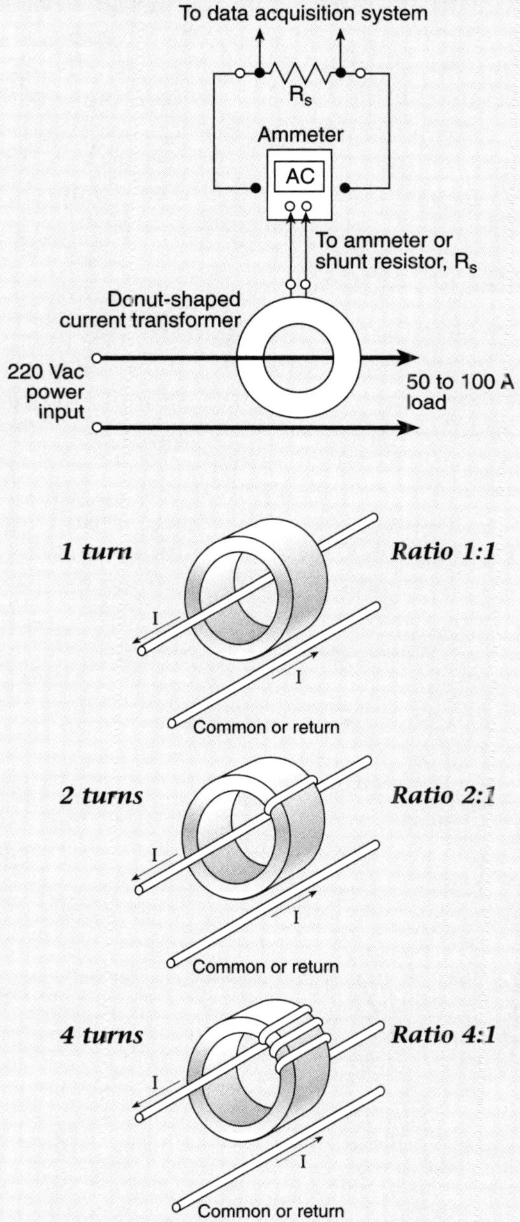

Fig. 5.27. Because it does not need a ground connection, a current transformer is isolated from both the input and output of the signal conditioner.

Isolated Sensor – Variable Reluctance/Magnetic

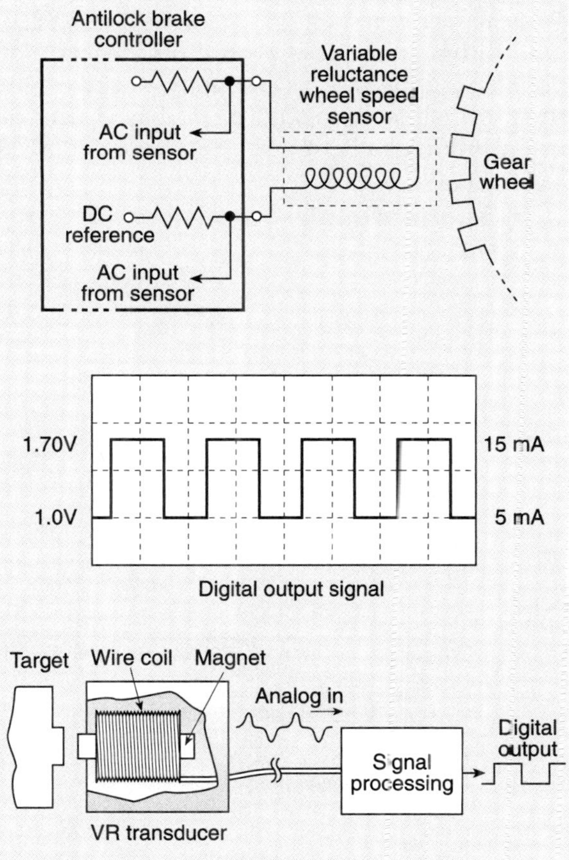

Fig. 5.28. Variable reluctance sensors comprise a coil of wire wound around a magnetic core. As a ferrous metal passes near one pole it disturbs the magnetic field and induces a small voltage in the coil. In this example, the voltage is amplified, shaped, and converted to a digital signal for indicating vehicle wheel speed.

mounting base if needed in some rare applications. (See Chapter 7, Strain Measurements.)

LVDTs contain a modulator and demodulator, either internally or externally, require some small dc power, and provide a small ac or dc signal to the data acquisition system. Often they are scaled to output 0 to 5 V. LVDTs can measure both position and acceleration. (See Chapter 9, Displacement and Position Sensing.)

Optical devices such as encoders are widely used in linear and rotary position sensors. The possible configurations are many, but basically, their principle of operation is based on the interruption of a light beam between an optical transmitter and receiver. A revolving opaque disc with multiple apertures placed between the transmitter and receiver alternately lets light through to generate pulses. Usually, LEDs generate the light and a photo diode on the opposite side detects the resulting pulses, which are then counted. The pulses can indicate position or velocity. (See Chapter 9, Displacement and Position Sensing.)

LINEARIZATION

Why Linearization Is Needed

The transfer function for many electronic devices, which relates the input to output, contains a nonlinear factor. In most cases this factor is small enough to be ignored. However, in some applications it must be compensated either in hardware or software. Thermocouples, for example, have a nonlinear relationship from input tempera-

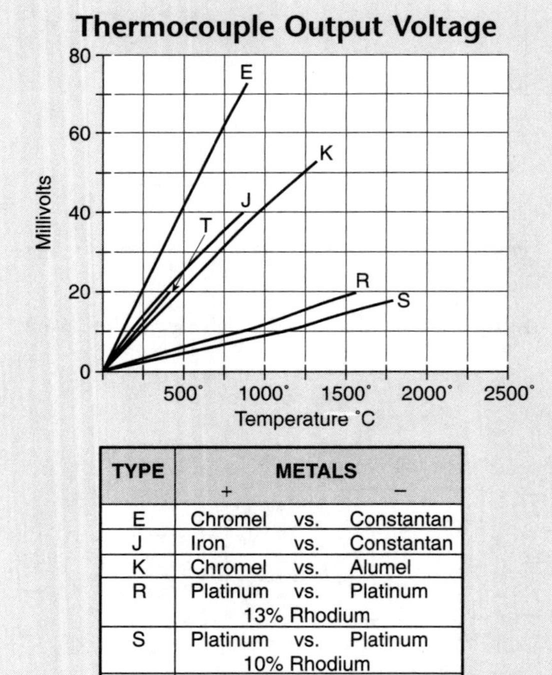

TYPE	METALS +	−	
E	Chromel	vs.	Constantan
J	Iron	vs.	Constantan
K	Chromel	vs.	Alumel
R	Platinum	vs.	Platinum 13% Rhodium
S	Platinum	vs.	Platinum 10% Rhodium
T	Copper	vs.	Constantan

Fig. 5.29. Although some thermocouples must be both thermally and electrically connected to the specimen under test, many may be purchased with insulated junctions, which isolates them from making high-voltage and ground-loop connections to the signal conditioner.

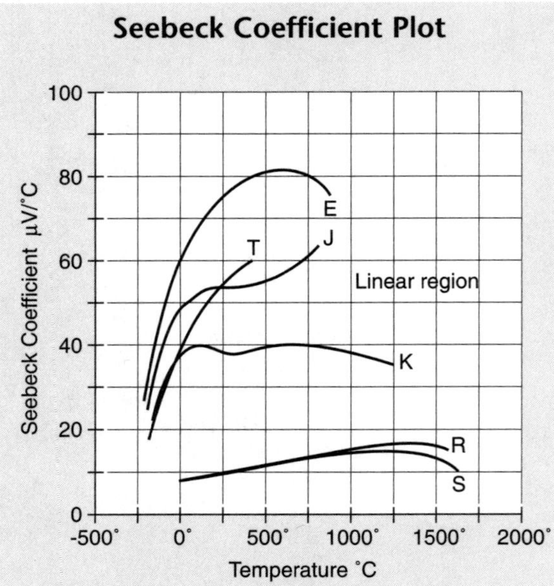

Fig. 5.30. The slope of the Seebeck coefficient plotted against temperature clearly illustrates that the thermocouple is a non-linear device.

ture to output voltage, severe enough to require compensation. Figure 5.29 shows the output voltage for several types of thermocouples plotted against temperature. Thermocouple output voltages are based on the Seebeck effect (discussed further in Chapter 6). When the slope of the Seebeck coefficient is plotted vs. temperature, the output response is clearly nonlinear as shown in Figure 5.30. A linear device, by comparison, would plot a straight horizontal line. Only the K type thermocouple most readily approaches a straight line in the range from about 0 to 1,000°C.

The curve shows that one constant scale factor is not sufficient over the entire temperature range for a particular thermocouple type to maintain adequate accuracy. Higher accuracy comes from reading the thermocouple voltage with a voltmeter and applying it to the thermocouple table from the National Institute of Standards and Technology as shown in Figure 5.31.

A computer-based data acquisition system, in contrast, automates the temperature conversion process using the thermocouple voltage reading and an algorithm to solve a polynomial equation. The equation describing this relationship is:

EQN 5.05. Temperature Polynomial

$$T = a_0 + a_1x + a_2x^2 + a_3x^3 + \ldots a_nx^n$$

Where: T = temperature, °C

x = thermocouple voltage, V

a = polynomial coefficients unique to each type of thermocouple

n = maximum order of the polynomial

The accuracy increases proportionally to the order of n. For example, when n = 9, a ±1°C accuracy may be realized. But because high order polynomials take time to process, lower orders may be used over limited temperature ranges to increase the processing speed.

Software Linearization

The polynomials are put to work in the data acquisition system's computer to calculate the real temperature for the thermocouple voltage. Typically, the computer program handles a nested polynomial to speed the process rather than compute the exponents directly. Nested

Temperature Conversion Table in °C (IPTS 1968)												
mV	.00	.01	.02	.03	.04	.05	.06	.07	.08	.09	.10	mV
0.00	0.00	0.17	0.34	0.51	0.68	0.85	1.02	1.19	1.36	1.53	1.70	0.00
0.10	1.70	1.87	2.04	2.21	2.38	2.55	2.72	2.89	3.06	3.23	3.40	0.10
0.20	3.40	3.57	3.74	3.91	4.08	4.25	4.42	4.58	4.75	4.92	5.09	0.20
0.30	5.09	5.26	5.43	5.60	5.77	5.94	6.11	6.27	6.44	6.61	6.78	0.30
0.40	6.78	6.95	7.12	7.29	7.46	7.62	7.79	7.96	8.13	8.30	8.47	0.40
0.50	8.47	8.63	8.80	8.97	9.14	9.31	9.47	9.64	9.81	9.98	10.15	0.50
0.60	10.15	10.31	10.48	10.65	10.82	10.98	11.15	11.32	11.49	11.65	11.82	0.60
0.70	11.82	11.99	12.16	12.32	12.49	12.66	12.83	12.99	13.16	13.33	13.49	0.70
0.80	13.49	13.66	13.83	13.99	14.16	14.33	14.49	14.66	14.83	14.99	15.16	0.80
0.90	15.16	15.33	15.49	15.66	15.83	15.99	16.16	16.33	16.49	16.66	16.83	0.90
1.00	16.83	16.99	17.16	17.32	17.49	17.66	17.82	17.99	18.15	18.32	18.48	1.00
1.10	18.48	18.65	18.82	18.98	19.15	19.31	19.48	19.64	19.81	19.97	20.14	1.10
1.20	20.14	20.13	20.47	20.64	20.80	20.97	21.13	21.30	21.46	21.63	21.79	1.20
1.30	21.79	21.96	22.12	22.29	22.45	22.62	22.78	22.94	23.11	23.27	23.44	1.30
1.40	23.44	23.60	23.77	23.93	24.10	24.26	24.42	24.59	24.75	24.92	25.08	1.40

Fig. 5.31. *Because thermocouple outputs are non-linear, a table is an accurate method for converting a voltage reading to temperature for a specific type of thermocouple.*

NIST Polynomial Coefficients						
	TYPE E	TYPE J	TYPE K	TYPE R	TYPE S	TYPE T
	Nickel-10% Chromium(+) Versus Constantan (–)	Iron (+) Versus Constantan (–)	Nickel-10% Chromium (+) Versus Nickel-5% (–) (Aluminum Silicon)	Platinum-13% Rhodium (+) Versus Platinum (–)	Platinum-10% Rhodium Versus Platinum (–)	Copper (+) Versus Constantan (–)
	-100°C to 1000°C ±0.5°C 9th order	0°C to 760°C ±0.1°C 5th order	0°C to 1370°C ±0.7°C 8th order	0°C to 1000°C ±0.5°C 8th order	0°C to 1750°C ±1°C 9th order	-160°C to 400°C ±0.5°C 7th order
a_0	0.104967248	-0.048868252	0.226584602	0.263632917	0.927763167	0.100860910
a_1	17189.45282	19873.14503	24152.10900	179075.491	169526.5150	25727.94369
a_2	-282639.0850	-218614.5353	67233.4248	-48840341.37	-31568363.94	-767345.8295
a_3	12695339.5	11569199.78	2210340.682	1.90002E + 10	8990730663	78025595.81
a_4	-448703084.6	-264917531.4	-860963914.9	-4.82704E + 12	-1.63565E + 12	-9247486589
a_5	1.10866E + 10	2018441314	4.83506E + 10	7.62091E + 14	1.88027E + 14	6.97688E + 11
a_6	-1.76807E + 11		-1.18452E + 12	-7.20026E + 16	-1.37241E + 16	-2.66192E + 13
a_7	1.71842E + 12		1.38690E + 13	3.71496E + 18	6.17501E + 17	3.94078E + 14
a_8	-9.19278E +12		-6.33708E + 13	-8.03104E + 19	-1.56105E + 19	
a_9	2.06132E + 13				1.69535E + 20	

Temperature Conversion Equation: $T = a_0 + a_1x + a_2x^2 + ... + a_nx^n$
Nested Polynomial Form: $T = a_0 + x(a_1 + x(a_2 + x(a_3 + x(a_4 + a_5x))))$ (5th order)

Fig. 5.32. The NIST polynomial table is a more accurate means of calculating a linearizing function for a particular thermocouple than a single coefficient, even though it uses a polynomial equation.

polynomials are the only practical way of dealing with complicated equations. Without such techniques it is difficult to handle large state tables that have more than a few hundred entries. (See Figure 5.32.) Also, high order polynomials can be computed faster when the thermocouple characteristic curve can be divided into several sectors and each sector approximated by a third order polynomial as shown in Figure 5.33.

Hardware Linearization

Hardware also may be designed to accommodate the nonlinearity of a thermocouple, but the circuitry becomes complex and expensive in order to reduce its susceptibility to errors from outside influences such as electrical noise and temperature variations within the circuits. The compensating circuitry is nonlinear, and contains breakpoints programmed with diodes, resistors, and reference voltages, all subject to errors that are avoided in software compensation methods. However, several modules are commercially available with excellent, stable, built-in linearizing circuits. The thermocouple voltage is extremely small and most signal conditioners concentrate less on compensation and more on amplifying the signal while rejecting common-mode noise. Alternative digital hardware methods use a look-up table to convert the thermocouple voltage to a corresponding temperature.

Curve Divided into Sectors

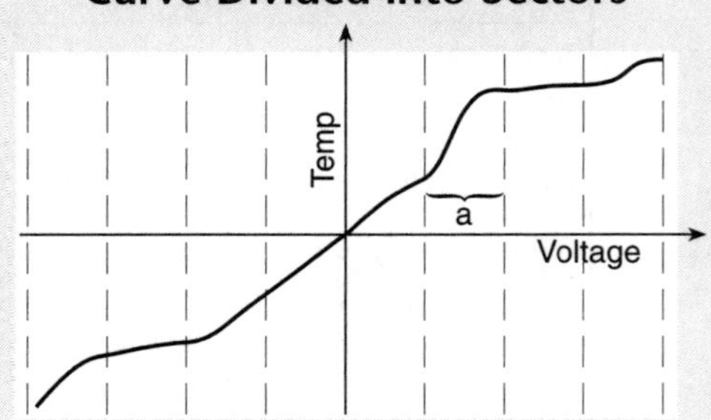

$T_a = bx + cx^2 + dx^3$

Fig. 5.33. Although a computer usually finds the solution to the NIST polynomial, breaking the curve into sections representing lower order polynomials can accelerate the process.

CIRCUIT PROTECTION

Hazards to Instrumentation Circuits

Many data acquisition systems contain solid-state multiplexing circuits to rapidly scan multiple input channels. Their inputs are typically limited to less than 30 V and may be damaged when exposed to higher voltages. Other solid-state devices in a measurement system including input amplifiers and bias sources also are limited to low voltages. However, these inputs can be protected with a programmable attenuator and isolation amplifiers that isolate the high voltage input stage from the solid-state circuitry. (See Figure 5.34.)

Another consideration often overlooked is connecting active inputs to an unpowered data acquisition system. Common safety practice calls for all signals connected to the input of the unpowered data acquisition system to be disconnected or their power removed. Frequently, de-energized data acquisition system signal conditioners have substantially lower input impedances than when energized, and even low voltage input signals higher than 0.5 Vdc can damage the signal conditioners' input circuits.

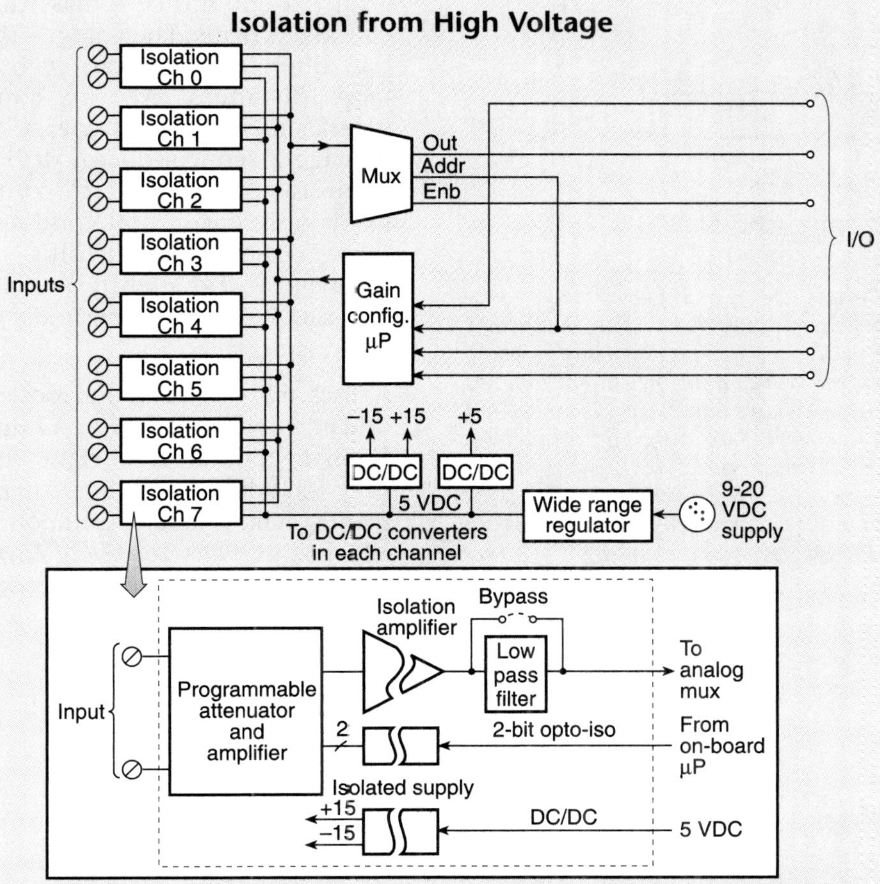

Fig. 5.34. The isolated voltage-input modules let data acquisition systems isolate several channels of analog input, up to 500 V channel-to-channel and channel-to-system.

Overload Protection

Several methods are used to protect signal conditioner inputs from damage when exposed to transients of 10 to 100 V. Often, a 1,000-Ω, current-limiting resistor is in series with the input when no voltage is applied to the instrument amplifiers' input. For transient input voltages to 3,000 V and above, a series resistor and a transient voltage suppressor are installed across the input terminals. (See Figure 5.35.)

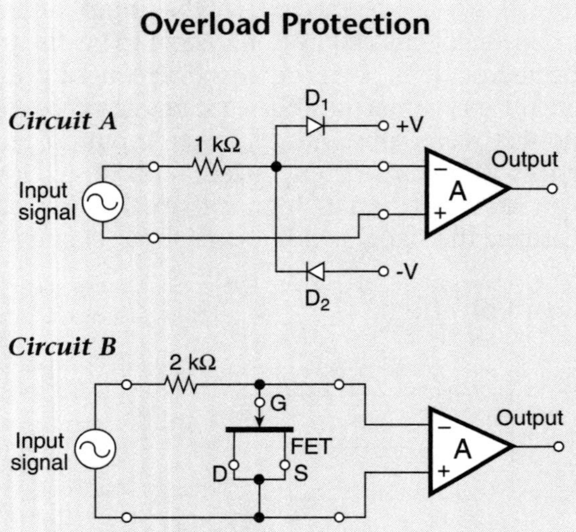

Fig. 5.35. Certain CMOS multiplexers may be protected against overvoltage destruction with diode and resistor networks that shut down any parasitic transistors in the device, limit the current input to a safe level, and shunt input signals to ground when the power supplies are turned off (Circuit A). JFET transistors also protect multiplexer inputs when they are connected as diodes. The JFET clamps at about 0.6 V, protecting the sensitive op amp input from destruction (Circuit B).

ESD Protection

Electrostatic discharge damage is a special problem that can be avoided by handling individual circuit boards carefully and providing proper shielding. ESD comes from the static charge accumulated on many different kinds of materials, which finds a return to ground or a mass that attracts the excess electrons. The charge can create potential differences that can eventually arc over large distances. An arc containing only several microjoules of energy can destroy or damage a semiconductor device. Grounding alone is not sufficient to control ESD build up, it only ensures all conductors are at the same potential. Controlling humidity to about 40% and slightly ionizing the air are the most effective methods of controlling static charge.

A discharge can travel one foot in one nanosecond and could rise to 5 A. A number of devices simulate conditions for static discharge protection, including a gun that generates pulses at a fixed voltage and rate. Component testing usually begins with relatively low voltages and gradually progresses to higher values.

Chapter 6

TEMPERATURE MEASUREMENT

THERMOCOUPLE BASICS

The Gradient Nature of Thermocouples

Thermocouples (TCs) are probably the most widely used and least understood of all temperature measuring devices. When connected in pairs, TCs are simple and efficient sensors that output an extremely small dc voltage proportional to the temperature difference between the two junctions in a closed thermoelectric circuit. (See Figure 6.01.) One junction is normally held at a constant reference temperature while the opposite junction is immersed in the environment to be measured. The principle of operation depends on the unique value of thermal emf measured between the open ends of the leads and the junction of two dissimilar metals held at a specific temperature. The principle is called the Seebeck Effect, named after the discoverer. The amount of voltage present at the open ends of the sensor and the range of temperatures the device can measure depend on the Seebeck coefficient, which in turn depends upon the chemical composition of the materials comprising the thermocouple wire. The Seebeck voltage is calculated from:

Equation 6.01. Seebeck Voltage

$$\Delta eAB = \alpha \Delta T$$

Where:

eAB = Seebeck voltage

T = temperature at the thermocouple junction

α = Seebeck coefficient

Δ = a small change in voltage corresponding to a small change in temperature

Thermocouple junctions alone do not generate voltages. The voltage or potential difference that develops at the output (open) end is a

Type T Basic Thermocouple Circuit

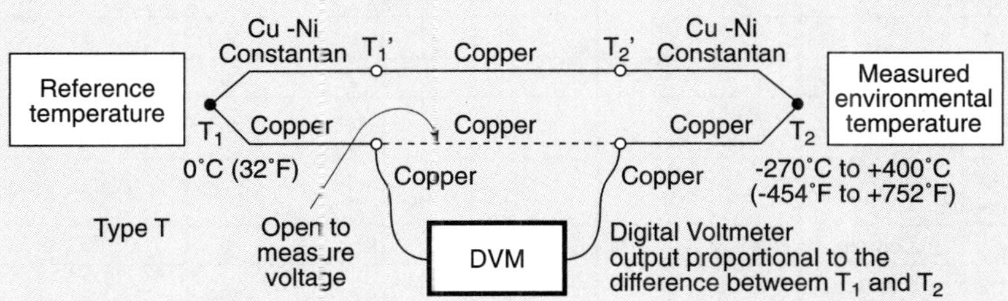

Fig. 6.01. A basic thermocouple measurement system requires two sensors, one for the environment under measurement and the other, a reference junction, normally held to 0°C (32°F). Type-T is one of the dozen or more common thermocouples frequently used in general-purpose temperature measuring applications. It is made of copper and constantan metals and typically operates from –270 to +400°C or –454 to +752°F.

function of both the temperature of the junction T_1 and the temperature of the open end T_1'. T_1' must be held at a constant temperature, such as 0°C, to ensure that the open end voltage changes in proportion to the temperature changes in T_1. In principle, a TC can be made from any two dissimilar metals such as nickel and iron. In practice, however, only a few TC types have become standard because their temperature coefficients are highly repeatable, they are rugged, and they output relatively large voltages. The most common thermocouple types are called J, K, T, and E, followed by N28, N14, S, R, and B. (See the table in Figure 6.02.) In theory, the junction temperature can be inferred from the Seebeck voltage by consulting standard tables. In practice, however, this voltage cannot be used directly because the thermocouple wire connection to the copper terminal at the measurement device itself constitutes a thermocouple junction (unless the TC lead is also copper) and outputs another emf that must be compensated.

Cold-Junction Compensation

A cold-reference-junction thermocouple immersed in an actual ice-water bath and connected in series with the measuring thermocouple is the classical method used to compensate the emf at the instrument terminals. (See Figure 6.03.) In this example, both copper leads connect to the instrument's input terminals. An alternative method uses a single thermocouple with the copper/constantan connection immersed in the reference ice water bath, also represented in Figure 6.03. The constantan/copper thermocouple junction J_2 in the ice bath contributes a small emf that subtracts from the emf from thermocouple J_1, so the voltage measured at the instrument or data acquisition system input terminals corresponds accurately to the NIST tables. Likewise,

Common Thermocouple Types

Type	Metal +	Metal –	Standard color code +	Standard color code –	Ω/double foot 20 AWG	Seebeck coefficient S(μV/°C)	@ T(°C)	°C standard wire error	NBS specified materials range* (°C)
B	Platinum-6% Rhodium	Platinum-30% Rhodium	–		0.2	6	600	4.4 to 8.6	0 to 1820**
E	Nickel-10% Chromium	Constantan	Violet	Red	0.71	58.5	0	1.7 to 4.4	-270 to 1000
J	Iron	Constantan	White	Red	0.36	50.2	0	1.1 to 2.9	-210 to 760
K	Nickel-10% Chromium	Nickel	Yellow	Red	0.59	39.4	0	1.1 to 2.9	-270 to 1372
N (AWG 14)	Nicrosil	Nisil	–		–	39	600	–	0 to 1300
N (AWG 28)	Nicrosil	Nisil	–		–	26.2	0	–	-270 to 400
R	Platinum-13% Rhodium	Platinum	–		0.19	11.5	600	1.4 to 3.8	-50 to 1768
S	Platinum-10% Rhodium	Platinum	–		0.19	10.3	600	1.4 to 3.8	-50 to 1768
T	Copper	Constantan	Blue	Red	0.30	38	0	0.8 to 2.9	-270 to 400
W-Re	Tungsten-5% Rhenium	Tungsten-26% Rhenium	–		–	19.5	600	–	0 to 2320

* Material range is for 8 AWG wire; decreases with deceasing wire size

** Type B double-valued below 42°C – curve fit specified only above 130°C

Fig. 6.02. NIST's (National Institute of Standards and Technology) thermocouple emf tables publish the emf output of a thermocouple based on a corresponding reference thermocouple junction held at 0°C.

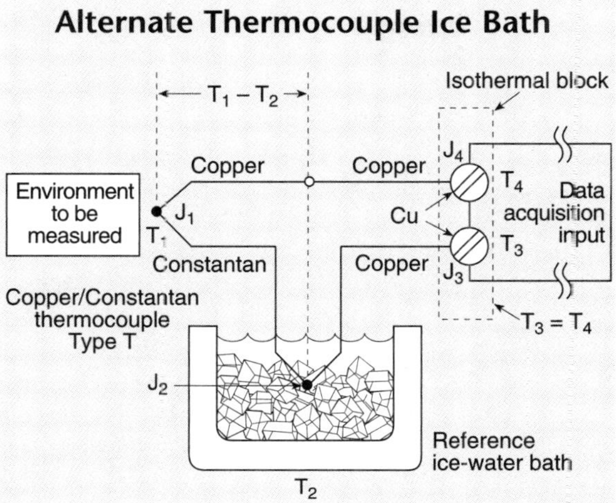

Fig. 6.03. Whether J2 is a purchased thermocouple or not, the junction formed by the constantan and copper lead wires at J2 must be placed in the ice bath for temperature compensation.

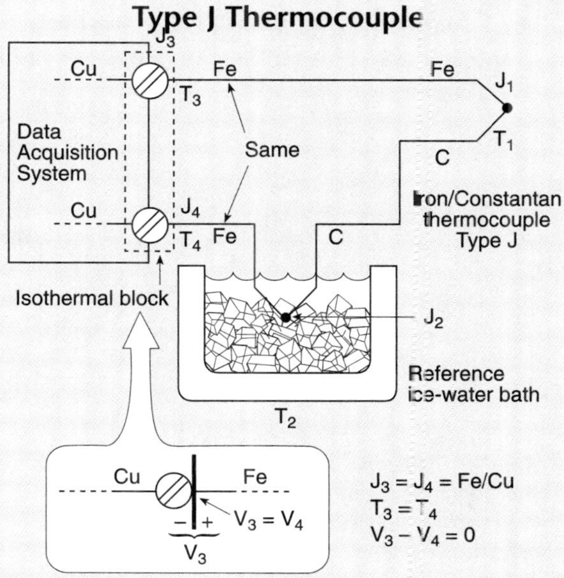

Fig. 6.04. One lead of a type-T thermocouple is copper, so it does not require temperature compensation when connecting to copper terminals. A Type J constantan/iron thermocouple, on the other hand, needs a closer look. J2 remains constant in the ice bath, and J1 measures the environment. Although J3 and J4 are effectively thermocouple junctions, they are at the same temperature on the isothermal block, so they generate equal and opposite voltages and cancel. Without an isothermal block, copper wire leads would be added between the input terminal and the copper/iron leads, and the copper/iron junctions (J3 and J4) would be held in an ice bath as well.

the copper wires connected to the copper terminals on the instrument's isothermal block do not need compensation because they are all copper at the same temperature. The voltage reading comes entirely from the NIST-adjusted constantan/copper thermocouple wire.

The above example is a special case, however, because one lead of the type-T thermocouple is copper. A constantan/iron thermocouple, on the other hand, needs further consideration. (See Figure 6.04.) Here, J2 in the ice bath is held constant, and J1 measures the environment. Although J3 and J4 are effectively thermocouple junctions, they are at the same temperature on the isothermal block, so they output equal and opposite voltages and thus cancel. The net voltage is then the thermocouple J1 output representing T1, calibrated to the NIST standard table. If the I/O block were not isothermal, copper wire leads would be added between the input terminal and the copper/iron leads, and the copper/iron junctions (J3 and J4) would be held in an ice bath as well, as illustrated in Figure 6.05.

Software Compensation

Ice baths and multiple reference junctions in large test fixtures are nuisances to set up and maintain, and fortunately they all can be eliminated. The ice bath can be ignored when the temperature of the lead wires and the reference junction points (isothermal terminal block at the instrument) are the same. The emf correction needed at the terminals can be referenced and compensated to the NIST standards through computer software.

When the ice baths are eliminated, cold junction compensation (CJC) is still necessary,

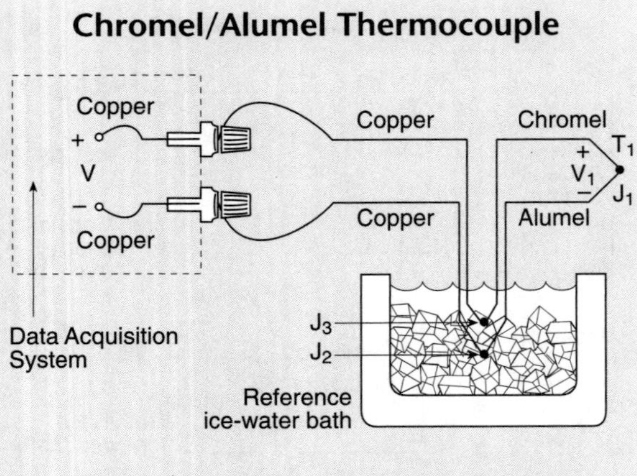

Fig. 6.05. The chromel and alumel wire connections at the copper leads constitute additional thermocouple junctions that must be held at the same, constant temperature. They generate equal and opposite potentials which prevent them from contributing to the voltage output from the chromel/alumel thermocouple.

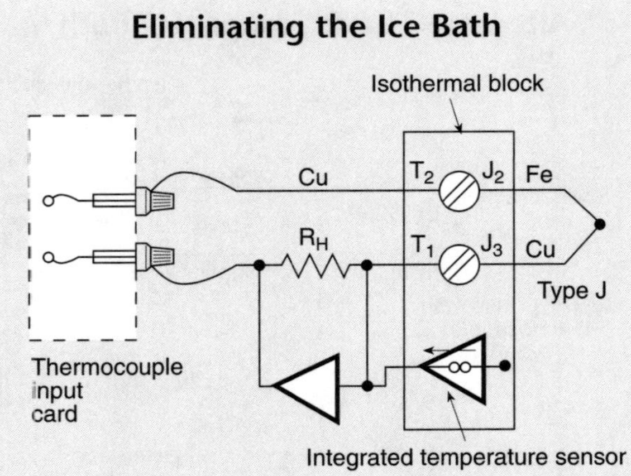

Fig. 6.06. A thermistor sensor placed near the lead wire connections is an alternative method of replacing the ice bath. The measured temperature is the difference between the thermocouple temperature and the reference thermistor temperature.

however, in order to obtain accurate thermocouple measurements. The software has to read the isothermal block temperature. One technique widely used is a thermistor, mounted close to the isothermal terminal block that connects to the external thermocouple leads. No temperature gradients are allowed in the region containing the thermistor and terminals. (See Figure 6.06.) The type of thermocouple employed is preprogrammed for its respective channel, and the dynamic input data for the software includes the isothermal block temperature and the measured environmental temperature. The software uses the isothermal block temperature and type of thermocouple to look up the value of the measured temperature corresponding to its voltage in a table, or it calculates the temperature more quickly with a polynomial equation. The method allows numerous channels of thermocouples of various types to be connected simultaneously while the computer handles all the conversions automatically.

Hardware Compensation

Although a polynomial approach is faster than a look-up table, a hardware method is even faster, because the correct voltage is immediately available to be scanned. One method is to insert a battery in the circuit to null the offset voltage from the reference junction so the net effect equals a 0°C junction. A more practical approach based on this principle is an "electronic ice point reference," which generates a compensating voltage as a function of the temperature sensing circuit powered by a battery or similar voltage source. (See Figure 6.07A.) The voltage then corresponds to an equivalent reference junction at 0°C.

Type Mixing

Thermocouple test systems often measure tens to hundreds of points simultaneously. In order to conveniently handle such large numbers of channels without the complication of separate, unique compensation TCs for each, thermo-

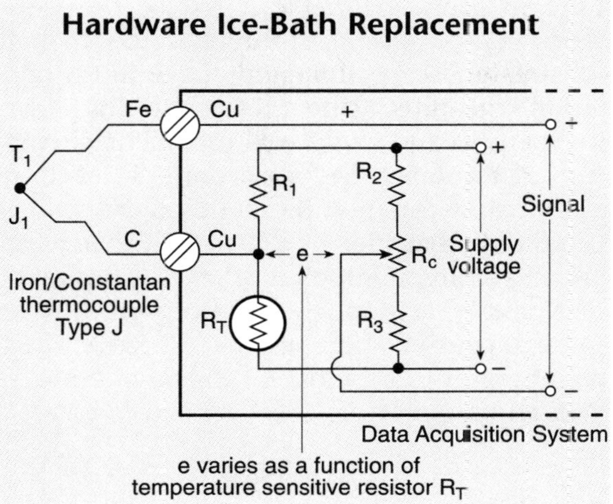

Fig. 6.07A. A number of electronic circuits or modules can replace the ice-bath. The temperature-sensitive resistor changes the calibrated value of voltage e in proportion to the amount of temperature compensation required.

couple-scanning modules come with multiple input channels and can accept any of the various types of thermocouples on any channel, simultaneously. They contain special copper-based input terminal blocks with numerous cold junction compensation sensors to ensure accurate readings, regardless of the sensor type used. Moreover, the module contains a built-in automatic zeroing channel as well as the cold-junction compensation channel. Although measurement speed is relatively slower than most other types of scanning modules, the readings are accurate, low noise, stable, and captured in only ms. For example, one TC channel can be measured in 3 ms, 14 TC channels in 16 ms, and up to 56 channels in 61 ms. Typical measurement accuracies are better than 0.7˚C, with channel-to-channel variation typically less than 0.5˚C. (See Figure 6.07B.)

Linearization

After setting up the equivalent ice point reference emf in either hardware or software, the

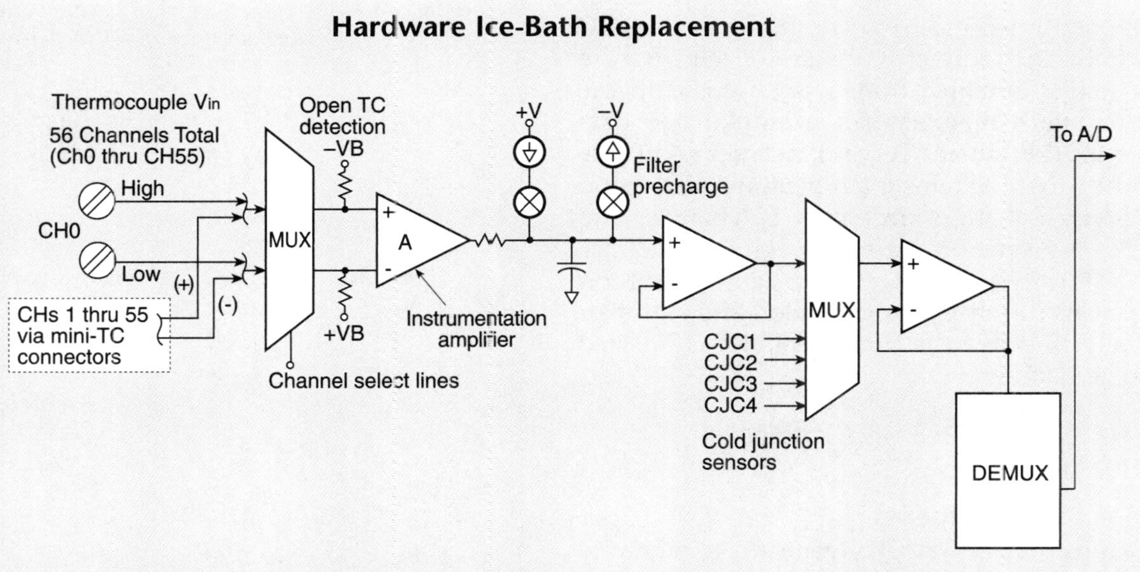

Fig. 6.07B. A typical input scanning module can accommodate up to 56 thermocouples of any type, and up to 896 channels can be connected to one analog-to-digital mainframe.

measured thermocouple output must be converted to a temperature reading. Thermocouple output is proportional to the temperature of the TC junction, but is not perfectly linear over a very wide range. (See Linearization, Chapter 5.)

The standard method for obtaining high conversion accuracy for any temperature uses the value of the measured thermocouple voltage plugged into a characteristic equation for that particular type thermocouple. The equation is a polynomial with an order of six to ten. The NIST table in Figure 5.32 (page 62) lists the polynomial coefficients for some common thermocouples. The computer automatically handles the calculation, but high-order polynomials take significant time to process. In order to accelerate the calculation, the thermocouple characteristic curve is divided into several segments. Each segment is then approximated by a lower order polynomial.

Analog circuits are employed occasionally to linearize the curves, but when the polynomial method is not used, the thermocouple output frequently connects to the input of an ADC where the correct voltage to temperature match is obtained from a table stored in the computer's memory. For example, one data acquisition system TC card includes a software driver that contains a temperature conversion library. It changes raw binary TC channels and CJC information into temperature readings. Some software packages for data acquisition systems supply CJC information and automatically linearize the thermocouples connected to the system.

THERMOCOUPLE MEASUREMENT PITFALLS

Noisy Environments

Because thermocouples generate a relatively small voltage, noise is always an issue. The most common source of noise is the utility power lines (50 or 60 Hz). Because thermocouple bandwidth is lower than 50 Hz, a simple filter in each channel can reduce the interfering ac line noise. Common filters include resistors and capacitors and active filters built around op amps. Although a passive RC filter is inexpensive and works well for analog circuits, it's not recommended for a multiplexed front end because the multiplexer's load can change the filter's characteristics. On the other hand, an active filter composed of an op amp and a few passive components works well, but it's more expensive and complex. Moreover, each channel must be calibrated to compensate for gain and offset errors. (See Figure 6.08.)

Passive Filters

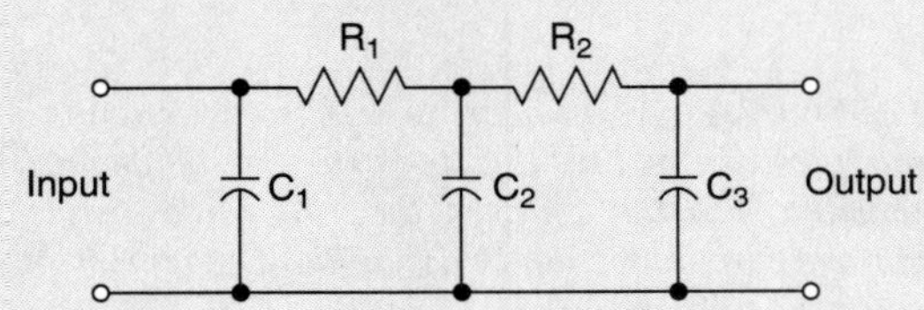

Fig. 6.08A. Passive filters come in a variety of configurations to suit the application. They are built in single or multiple sections to provide increasingly steeper slopes for faster roll-off.

Active Filters

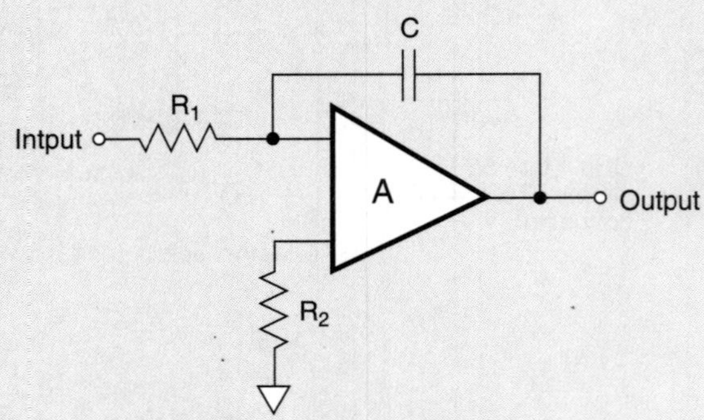

Fig. 6.08B. An active filter easily eliminates the most common sources of electrical noise that competes with the thermocouple signal such as the interference from 50/60 Hz supply lines.

Additional Concerns

Thermocouple Assembly

Thermocouples are twisted pairs of dissimilar wires that are soldered or welded together at the junction. When not assembled properly, they can produce a variety of errors. For example, wires should not be twisted together to form a junction; they should be soldered or welded. However, solder is sufficient only at relatively low temperatures, usually less than 200°C. And although soldering also introduces a third metal, such as a lead/tin alloy, it will not likely introduce errors if both sides of the junction are at the same temperature. Welding the junction is preferred, but it must be done without changing the wires' characteristics. Commercially manufactured thermocouple junctions are typically joined with capacitive discharge welders that ensure uniformity and prevent contamination.

Thermocouples can become un-calibrated and indicate the wrong temperature when the physical makeup of the wire is altered. Then it cannot meet the NIST standards. The change can come from a variety of sources, including exposure to temperature extremes, cold working the metal, stress placed on the cable when installed, vibration, or temperature gradients.

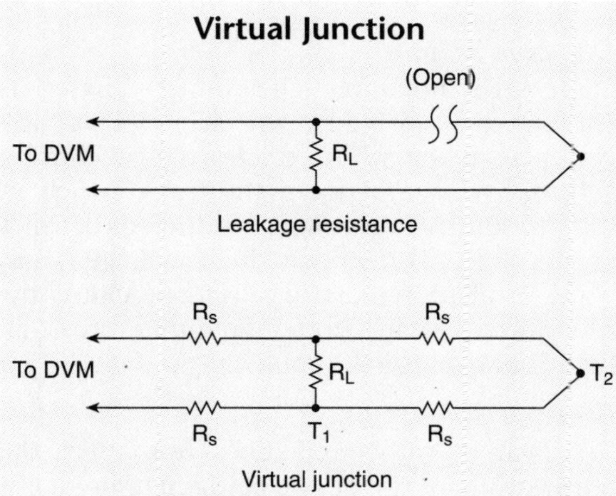

Fig. 6.09. A short circuit or an insulation failure between the leads of a thermocouple can form an unwanted, inadvertent thermocouple junction called a virtual junction.

The output of the thermocouple also can change when its insulation resistance decreases as the temperature increases. The change is exponential and can produce a leakage resistance so low that it bypasses an open-thermocouple wire detector circuit. In high-temperature applications using thin thermocouple wire, the insulation can degrade to the point of forming a virtual junction as illustrated in Figure 6.09. The data acquisition system will then measure the output voltage of the virtual junction at T_1 instead of the true junction at T_2.

In addition, high temperatures can release impurities and chemicals within the thermocouple wire insulation that diffuse into the thermocouple metal and change its characteristics. Then, the temperature vs. voltage relationship deviates from the published values. Choose protective insulation intended for high-temperature operation to minimize these problems.

Thermocouple Isolation

Thermocouple isolation reduces noise and errors typically introduced by ground loops. This is especially troublesome where numerous thermocouples with long leads fasten directly between an engine block (or another large metal object) and the thermocouple-measurement instrument. They may reference different grounds, and without isolation, the ground loop can introduce relatively large errors in the readings.

Auto-Zero Correction

Subtracting the output of a shorted channel from the measurement channel's readings can minimize the effects of time and temperature drift on the system's analog circuitry. Although extremely small, this drift can become a significant part of the low-level voltage supplied by a thermocouple.

One effective method of subtracting the offset due to drift is done in two steps. First, the internal channel sequencer switches to a reference node and stores the offset error voltage on a capacitor. Next, as the thermocouple channel switches onto the analog path, the stored error voltage is applied to the offset correction input of a differential amplifier and automatically nulls out the offset. (See Figure 6.10.)

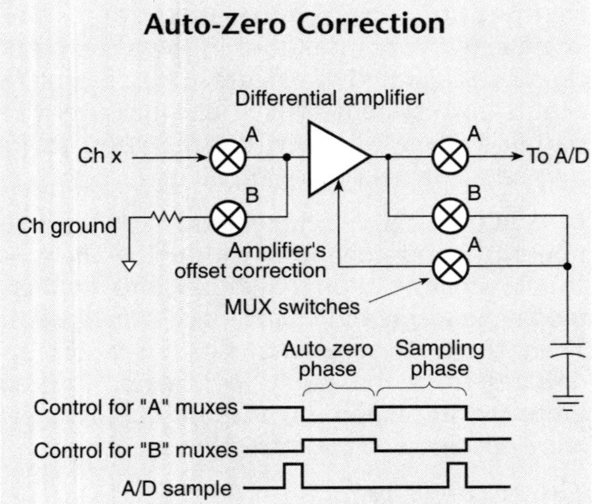

Fig. 6.10. Auto-Zero Correction compensates for analog circuitry drift over time and temperature. Although small, the offset could approach the magnitude of the thermocouple signal.

Open Thermocouple Detection

Detecting open thermocouples easily and quickly is especially critical in systems with numerous channels. Thermocouples tend to break or increase resistance when exposed to vibration, poor handling, and long service time. A simple open-thermocouple detection circuit comprises a small capacitor placed across the thermocouple leads and driven with a low-level current. The low impedance of the intact thermocouple presents a virtual short circuit to the capacitor so it cannot charge. When a thermocouple opens or significantly changes resistance, the capacitor charges and drives the input to one of the voltage rails, which indicates a defective thermocouple. (See Figure 6.11.)

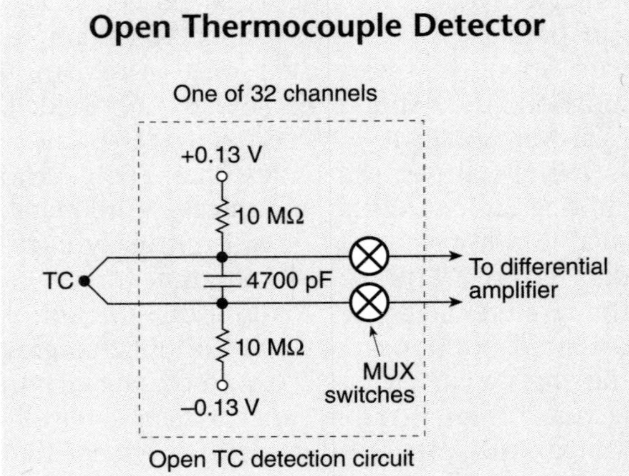

Fig. 6.11. The thermocouple provides a short-circuit path for dc around the capacitor, preventing it from charging through the resistors. When the thermocouple opens, due to rough handling or vibration, the capacitor charges and drives the input amplifier to the power supply rails, signaling a failure.

Galvanic Action

Some thermocouple insulating materials contain dyes that form an electrolyte in the presence of water. The electrolyte generates a galvanic voltage between the leads, which in turn, produces output signals hundreds of times greater than the net open-circuit voltage. Thus, good installation practice calls for shielding the thermocouple wires from high humidity and all liquids to avoid such problems.

Thermal Shunting

An ideal thermocouple does not affect the temperature of the device being measured, but a real thermocouple has mass that when added to the device under test can alter the temperature measurement. Thermocouple mass can be minimized with small diameter wires, but smaller wire is more susceptible to contamination, annealing, strain, and shunt impedance. One solution to help ease this problem is to use the small thermocouple wire at the junction but add special, heavier thermocouple extension wire to cover long distances. The material used in these extension wires has net open-circuit voltage coefficients similar to specific thermocouple types. Its series resistance is relatively low over long distances, and it can be pulled through conduit easier than premium grade thermocouple wire. In addition to its practical size advantage, extension wire is less expensive than standard thermocouple wire, especially platinum.

Despite these advantages, extension wire generally operates over a much narrower temperature range and is more likely to receive mechanical stress. For these reasons, the temperature gradient across the extension wire should be kept to a minimum to ensure accurate temperature measurements.

Improving Wire Calibration Accuracy

Thermocouple wire is manufactured to NIST specifications. Often, these specifications can be met more accurately when the wire is calibrated on site against a known temperature standard.

RTD MEASUREMENTS

Basics of Resistance Temperature Detectors

RTDs are composed of metals with a high positive temperature coefficient of resistance. Most RTDs are simply wire-wound or thin film resistors made of material with a known resistance vs. temperature relationship. Platinum is one of the most widely used materials for RTDs. They come in a wide range of accuracies, and the most accurate are also used as NIST temperature standards.

Platinum RTD resistances range from about 10 Ω for a birdcage configuration to 10 kΩ for a film type, but the most common is 100 Ω at 0°C. Commercial platinum wire has a standard temperature coefficient, α, of 0.00385 Ω/Ω/°C, and chemically pure platinum has a coefficient of 0.00392 Ω/Ω/°C.

The following equation shows the relationship between the sensor's relative change in resistance with a change in temperature at a specific α and nominal sensor resistance.

Equation 6.02. RTD Temperature Coefficient

$$\Delta R = \alpha R_O \Delta T,$$

Where:

α = temperature coefficient, Ω/Ω/°C

R_O = nominal sensor resistance at 0°C, Ω

ΔT = change in temperature from 0°C, C

A nominal 100-Ω platinum wire at 0°C will change resistance, either plus or minus, over a slope of 0.385 Ω/°C. For example, a 10°C rise in temperature will change the output of the sensor from 100 Ω to 103.85 Ω, and a 10°C fall in temperature will change the RTD resistance to 96.15 Ω.

Because RTD sensor resistances and temperature coefficients are relatively small, lead wires with a resistance as low as ten ohms and relatively high temperature coefficients can change the data acquisition system's calibration. The lead wire's resistance change over temperature can add to or subtract from the RTD sensor's output and produce appreciable errors in temperature measurement.

The resistance of the RTD (or any resistor) is determined by passing a measured current through it from a known voltage source. The resistance is then calculated using Ohms Law. To eliminate the measurement error contributed by lead wires, a second set of voltage sensing leads should be connected to the sensor's terminals and the opposite ends connected to corresponding sense terminals at the signal conditioner. This is called a four-wire RTD measurement. The sensor voltage is measured directly and eliminates the voltage drop in the current carrying leads.

Measurement Approaches

Configurations: 2, 3 and 4-wire

Five types of circuits are used for RTD measurements using two, three, and four lead wires: Two-wire with current source, four wire with current source, three-wire with current source, four-wire with voltage source, and three-wire with voltage source.

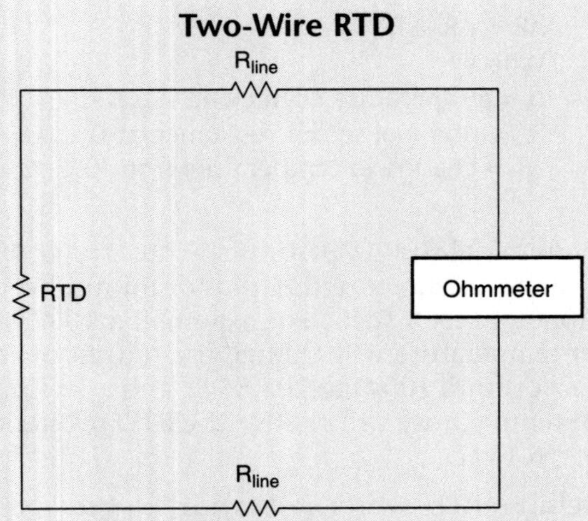

Fig. 6.12. The simplest arrangement for an RTD measurement is a simple series circuit containing only two wires connected to an ohmmeter.

Figure 6.12 shows a basic two-wire resistance measurement method. The RTD resistance is measured directly from the Ohmmeter. But this connection is rarely used since the lead wire resistance and temperature coefficient must be known. Often, both properties are not known, nor are they convenient to measure when setting up a test.

Figure 6.13 shows a basic four-wire measurement method using a current source. The RTD resistance is V/A. This connection is more accurate than the two-wire method, but it requires a high stability current source and four lead wires. Because the high-impedance voltmeter does not draw appreciable current, the voltage across the RDT equals V_m.

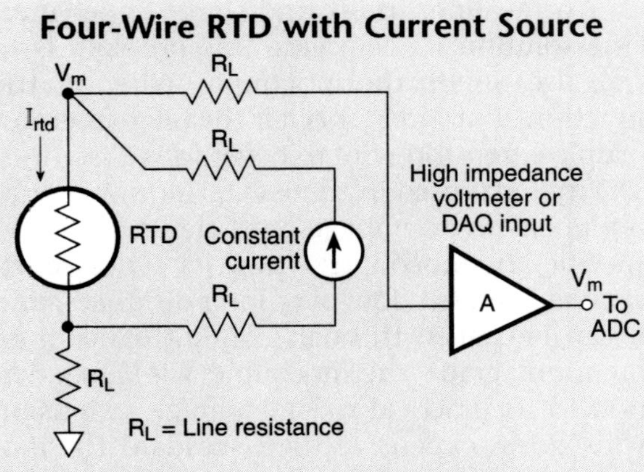

Fig. 6.13. The four-wire RTD method with a current supply eliminates the lead wire resistance as a source of error.

Equation 6.03. 4-Wire RTD With Current Source

$$R_{rtd} = \frac{V_m}{I_{rtd}}$$

Where:

R_{rtd} = RTD resistance, Ω

V_m = Voltmeter reading, V

I_{rtd} = RTD current, A

Figure 6.14 shows a three-wire measurement technique using a current source. The symbols

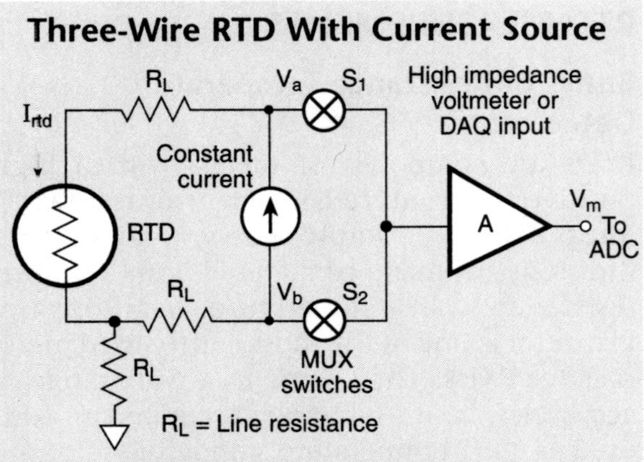

Fig. 6.14. The three-wire RTD method with a current supply is similar to the four-wire method. It simply eliminates one additional wire. Measure V_a first, then measure V_b.

V_a and V_b represent two voltages measured by the high-impedance voltmeter in sequence through switches (or a MUX), S_1 and S_2. The RTD resistance is derived from Kirchhoff's voltage law and by solving two simultaneous equations. (Illustrating the solution is beyond the scope of this book.) The benefit of this connection over that shown in Figure 6.13 is one less lead wire. However, this connection assumes that the two current-carrying wires have the same resistance.

Equation 6.04. 3-Wire RTD With Current Source

$$R_{rtd} = \frac{V_a - V_b}{I_{rtd}}$$

Figure 6.15 shows a four-wire measurement system using a voltage source. The RTD resistance also is derived from Kirchhoff's voltage law and four simultaneous equations based on the four voltages, V_a through V_d. The voltage source in this circuit can vary somewhat as long as the sense resistor remains stable.

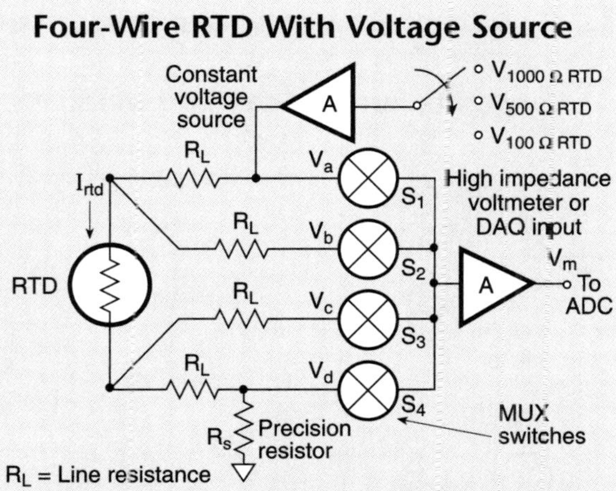

Fig. 6.15. The four-wire RTD circuit with a voltage source is more complex than the four-wire with current source, but the voltage is allowed to vary somewhat provided the sense resistor is stable.

Equation 6.05. 4-Wire RTD With Voltage Source

$$R_{rtd} = \frac{R_s (V_b - V_c)}{V_d}$$

Figure 6.16 shows a three-wire measurement technique using a voltage source. The RTD resistance is derived from Kirchhoff's voltage law and three simultaneous equations. The voltage source can vary as long as the sense resistor remains stable, and the circuit is accurate as long as the resistances of the two current-carrying wires are the same.

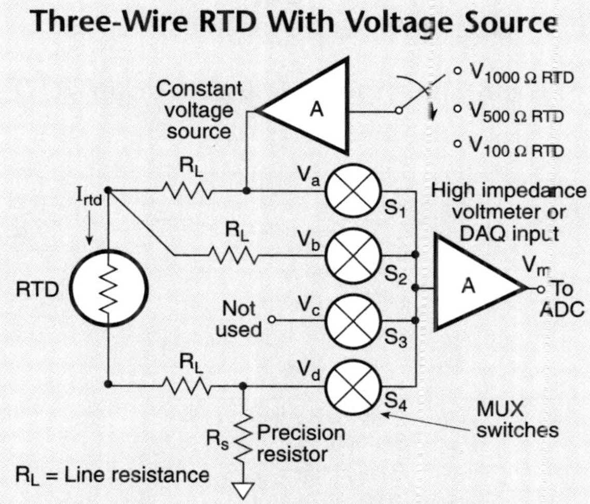

Fig. 6.16. This is a variation of the four-wire circuit with a voltage source and a stable sense resistor.

Equation 6.06. 3-Wire RTD With Voltage Source

$$R_{rtd} = \frac{R_s (2V_b - V_a - V_d)}{V_d}$$

The RTD output is more linear than the thermocouple, but its range is smaller. The Callendar-Van Dusen equation is often used to calculate the RTD resistance:

Equation 6.07. RTD Curve Fitting

$$R_T = R_o + R_o \alpha \left[T - \delta\left(\frac{T}{100} - 1\right)\left(\frac{T}{100}\right) - \beta\left(\frac{T}{100} - 1\right)\left(\frac{T^3}{100}\right)\right]$$

Where:

R_T = resistance at T, Ω

R_o = resistance at T = 0°C, Ω

α = temperature coefficient at T = 0°C

δ = 1.49 (for platinum)

β = 0, when T>0;

β = 0.11 when T<0

An alternative method involves measuring RTD resistances at four temperatures and solving a 20th order polynomial equation with these values. It provides more precise data than does the α, δ, and β coefficients in the Callendar-Van Dusen equation. The plot of the polynomial equation in Figure 6.17 shows the RTD to be more linear than the thermocouple when used below 800°C, the maximum temperature for RTDs.

Type S Thermocouple vs. Platinum RTD

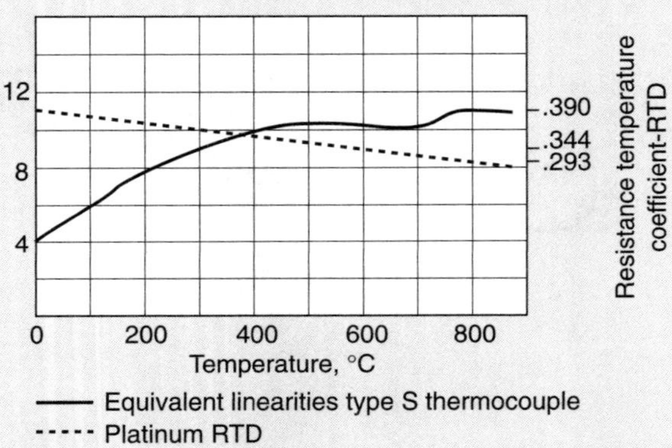

Fig. 6.17. Platinum is the material of choice for RTDs and thermocouples because it is stable and resists corrosion. Here, the RTD is shown to be more linear under temperatures of 800°C than the thermocouple.

Self-Heating

Another source of error in RTD measurements is resistive heating. The current, I, passing through the RTD sensor, R, dissipates power, $P = I^2R$. For example, 1 mA through a 100 Ω RTD generates 100 μW. This may seem insignificant, but it can raise the temperature of some RTDs a significant fraction of a degree. A typical RTD can change 1°C/mW by self-heating. When selecting smaller RTDs for faster response times, consider that they also can have larger self-heating errors.

A typical value for self-heating error is 1°C/mW in free air. An RTD immersed in a thermally conductive medium distributes this heat to the medium and the resulting error is smaller. The same RTD rises 0.1°C/mW in air flowing at one m/s. Using the minimum excitation current that provides the desired resolution, and using the largest physically practical RTD will help reduce self-heating errors.

Scanning Inputs

Because lower currents generate less heat, currents between 100 and 500 μA are typically used. This lowers the power dissipation to 10 to 25 μW, which most applications can tolerate. Further reducing the current lowers accuracy because they become more susceptible to noise and are more difficult to measure. But switching the current on only when the measurement is made can reduce the RTD's heat to below 10 μW. In a multichannel system, for example, the excitation current can be multiplexed, much like the analog inputs. In a 16-channel system, the current will

Constant-Current Scanning Module

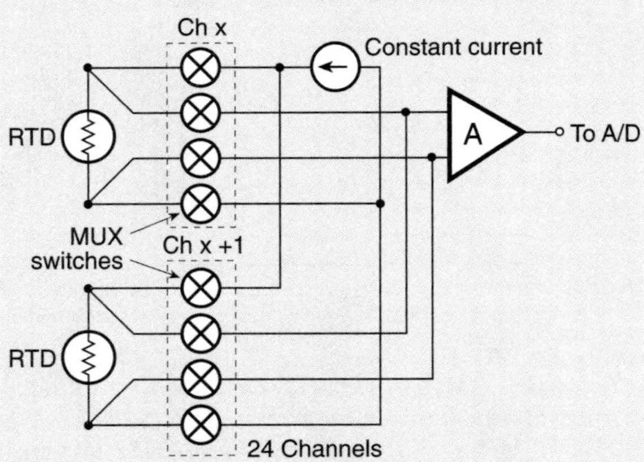

Fig. 6.18. The constant-current source is sequentially switched among the various RTD sensors to keep them cooler over the measurement interval and prevent resistive-heating errors.

only excite each RTD 1/16th of the time, reducing the power delivered to each RTD from 100% to only 6%.

Two practical methods for scanning an RTD include constant current and ratiometric. An example of a constant current circuit is shown in Figure 6.18. It's an RTD scanning module, which switches a single 500 µA constant current source among 16 channels. A series of front-end multiplexers direct the current to each channel sequentially while the measurement is being taken. Both three and four wire connections are supported to accommodate both types of RTDs. By applying current to one RTD at a time, errors due to resistive heating become negligible. Advantages of the constant current method include simple circuits and noise immunity. But the disadvantage is the high cost of buying or building an extremely stable constant current source.

Ratiometric Four-Wire RTD

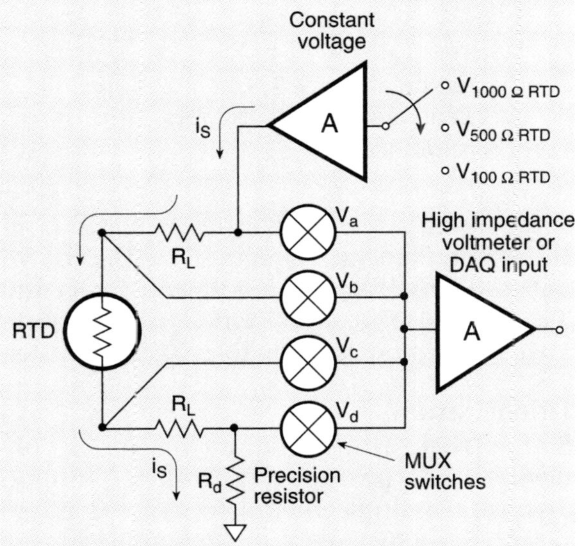

Fig. 6.19. Four voltage readings are taken for each RTD channel. The precision resistor measures I_S, the RTD current; V_b and V_c measure the RTD voltage; and the RTD resistance equals $(V_b - V_c)/I_S$.

By contrast, the ratiometric method uses a constant voltage source to provide a current, I_S, through the RTD and a resistor, R_d. Four voltage readings are taken for each RTD channel, V_a, V_b, V_c, and V_d. (See Figure 6.19.)

The current, voltage, and resistance of the RTD is:

Equation 6.08. 4-Wire RTD Ratiometric Measurement

$$I_S = \frac{V_d}{R_d}$$

$$V_{rtd} = V_b - V_c$$

$$R_{rtd} = \frac{V_{rtd}}{I_S}$$

For a three-wire connection (Figure 6.20), the voltage, $V_a - V_c$, includes the voltage drop across only one lead. Because the two extension wires to the transducer are made of the same metal, assume that the drop in the first wire is equal to the drop in the second wire. Therefore, the voltage across the RTD and its resistance is:

Equation 6.09. 3-Wire RTD Ratiometric Measurement

$$V_{rtd} = V_a - 2(V_a - V_b) - V_d$$

$$R_{rtd} = R_d\left(\frac{V_{rtd}}{V_d}\right)$$

Practical Precautions

RTDs require the same precautions that apply to thermocouples, including using shields and twisted-pair wire, proper sheathing, avoiding stress and steep gradients, and using large diameter extension wire. In addition, the RTD is more fragile than the thermocouple and needs to be protected during use. Also, thermal shunting is a bigger concern for RTDs than for thermocouples because the mass of the RTD is generally much larger. (See Figure 6.21.)

Ratiometric Three-Wire RTD

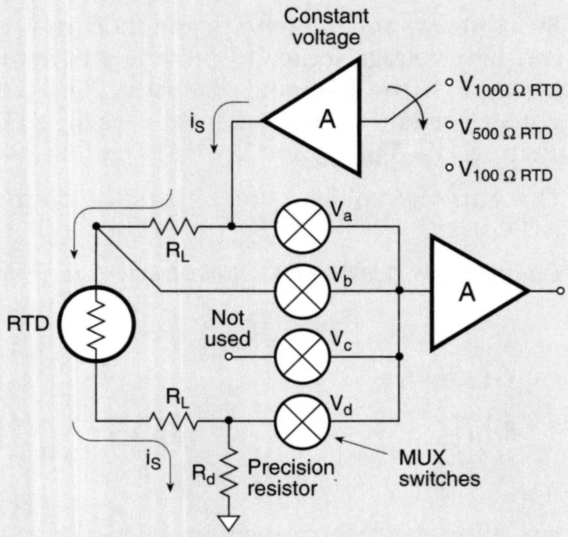

Fig. 6.20. The three-wire ratiometric circuit assumes that both sense-wire resistances in the four-wire circuit are the same. The equation for calculating RTD resistance simply accounts for it with a factor of two.

RTD Resistance Comparison: Small Resistance vs. Large Resistance		
	Small RTD	**Large RTD**
Response time	Fast	Slow
Thermal shunting	Low	Poor
Self-heating error	High	Low

Fig. 6.21. Although smaller RTDs respond faster to temperature changes, they are more susceptible to inaccuracy from self-heating.

THERMISTOR MEASUREMENTS

Basics of Thermistors

Thermistors are similar to RTDs in that they also change resistance between their terminals with a change in temperature. However, they can be made with either a positive or negative temperature coefficient. In addition, they have a much higher ratio of resistance change per °C (several %) than RTDs, which makes them more sensitive.

Thermistors are generally composed of semiconductor materials or oxides of common elements such as cobalt, copper, iron, manganese, magnesium, nickel, and others. They typically come with 3 to 6-in. leads, encapsulated, and color-coded. They are available in a range of accuracies from ±15°C to ±1°C, with a nominal resistance ranging from 2,000 to 10,000 Ω at 25°C. A value of 2252 Ω is common and can be used with most instruments. A plot of the temperature vs. resistance characteristic curves is usually provided with the device to determine the temperature from a known resistance. However, the devices are highly non-linear and the following equation may be used to calculate the temperature:

Equation 6.10. Thermistor Temperature

$$\frac{1}{T} = A + B(\log_e R) + C(\log_e R)^3$$

Where:
T = temperature, °K
A, B, and C = fitting constants
R = resistance, Ω

The constants A, B, and C are calculated from three simultaneous equations with known data sets: Insert R1 and T1; R2 and T2; R3 and T3, then solve for A, B, and C. Interpolation yields a solution accurate to ±0.01°C or better.

Linearization

Some thermistor manufacturers supply devices that provide a near-linear output. They use multiple thermistors (positive and negative coefficients) or a combination of thermistors and metal film resistors in a single package. When connected in certain networks, they produce a linearly varying voltage or resistance proportional to temperature. A widely used equation for the voltage divider shown in Figure 6.22 is:

Equation 6.11. Thermistor Voltage Divider

$$E_{out} = E_{in}\left(\frac{R}{R + R_o}\right)$$

Where:

E_{out} is the voltage drop across R

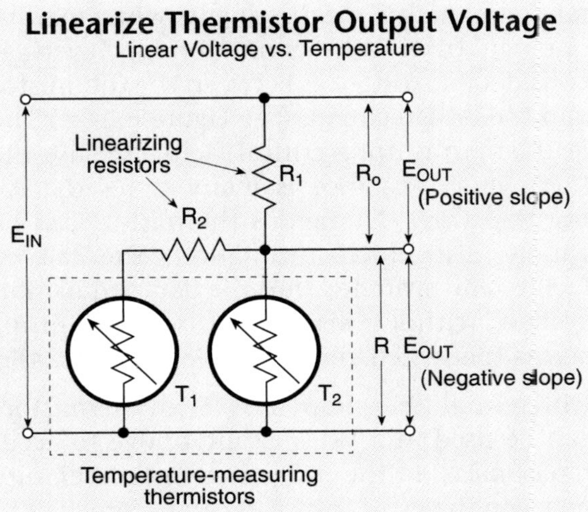

Fig. 6.22. The compensating resistors in series with the thermistors improve linearity near the center of the thermistor's S-shaped characteristic curve. This is where the sensitivity is the greatest, and its operating temperatures can be extended to cover a wider range.

If R is a thermistor, and the output voltage is plotted against the temperature, the curve resembles an S-shape with a fairly straight center portion. However, adding other resistors or thermistors to R linearizes the center portion of the curve over a wider temperature range. The linear section follows the equation of a straight line, Y = mX +b:

For the voltage mode:

Equation 6.12. Thermistor Voltage Mode

$E_{out} = \pm MT + b$

Where:

T = temperature in °C or °F

b = value of E_{out} when T = 0

M = slope, volts per degree T in °C or °F, V/°C or V/°F

For the resistance mode, see Figure 6.23:

Equation 6.13. Thermistor Resistance Mode

$R_t = MT + b$

Where:

T = temperature in °C or °F

b = value of the total network resistance R_t in Ω when T = 0

M = slope, Ω per degree T in °C or °F, Ω/°C or Ω/°F

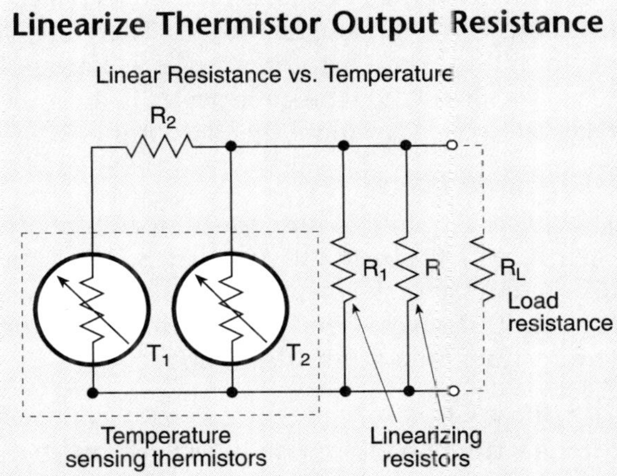

Fig. 6.23. Compensating resistors in the network linearize the resistance change vs. temperature in the same manner as they do for the voltage mode.

Although a lot of research has gone into developing linear thermistors, most modern data acquisition system controllers and software handle the linearization, which makes hardware linearization methods virtually obsolete.

Stability

Thermistors are inherently and reasonably stable devices, not normally subject to large changes in nominal resistance with aging, nor with exposure to strong radiation fields. However, prolonged operation over 90°C can change the tolerance of thermistors, particularly those with values less than 2,000 Ω. They are smaller and more fragile than thermocouples and RTDs, so they cannot tolerate much mishandling.

Time Constant

The time required for a thermistor to reach 63% of its final resistance value after being thrust into a new temperature environment is called its time constant. The time constant for an unprotected thermistor placed in a liquid bath may range from 1 to 2.5 sec. The same device exposed to an air environment might require 10 sec, while an insulated unit could require up to 25 sec. Seven time constants is a universally accepted value to consider when the device has reached its plateau or about 99% of its final value. Therefore, a device in the liquid bath might take as long as 7 sec to stabilize, while the same device in air could take 125 seconds or more than two minutes.

Dissipation Factor

The power required to raise the temperature of a thermistor 1°C above the ambient is called the dissipation factor. It is typically in the mW range for most devices. The maximum operating temperature for a thermistor is about 150°C.

Tolerance Curves

Manufacturers have not standardized on thermistor characteristic curves to the extent they have for thermocouples and RTDs. Thermistors are well suited to measuring temperature set points, and each thermistor brand comes with its unique curve which is often used to design ON/OFF control circuits.

Measurement Approaches

Temperature Measurement

Wheatstone bridge: Thermistors provide accurate temperature measurements when used in one leg of a Wheatstone bridge, even at considerable distances between the thermistor and the bridge circuit. (See Figure 6.24A.) The lead length is not a critical factor because the thermistor resistance is many times that of the lead wires. Numerous thermistors can be widely distributed throughout the lab or facility and switched into the data acquisition system without significant voltage drops across the switch contacts. (See Figure 6.24B.)

Differential thermometers: Two thermistors can be used in a Wheatstone bridge to accurately measure the difference in temperature between them. Thermistors can be attached to any heat conducting medium in a system at various points to measure the temperature gradient along its length. Two or more thermistors may be placed in a room to measure temperatures at several different elevations using the same basic switching arrangement.

Thermistors in a Wheatstone Bridge

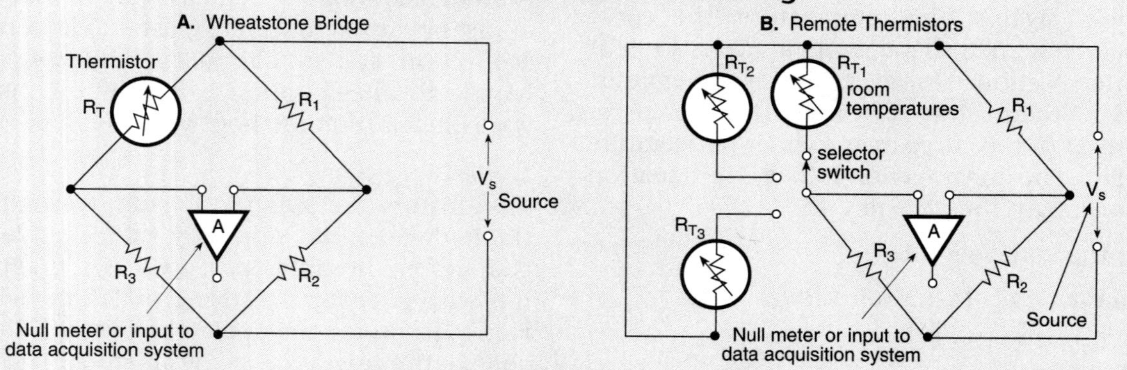

***Fig. 6.24.** An accurate temperature sensor can be fashioned from a thermistor in one leg of a bridge circuit. Lead length is not significant; so several sensors may be switched in and out of a single monitor without loosing accuracy. Two thermistors make a differential thermometer that can be used for measuring temperature changes along a piping system or between various elevations in a building to balance the heating and air conditioning unit.*

Chapter 7

STRAIN MEASUREMENTS

STRAIN BASICS

Strain gages are sensing devices used in a variety of physical test and measurement applications. They change resistance at their output terminals when stretched or compressed. Because of this characteristic, the gages are typically bonded to the surface of a solid material and measure its minute dimensional changes when put in compression or tension. Strain gages and strain gage principles are often used in devices for measuring acceleration, pressure, tension, and force. Strain is a dimensionless unit, defined as a change in length per unit length. For example, if a 1-m long bar stretches to 1.000002 m, the strain is defined as 2 microstrains. Strain gages have a characteristic gage factor, defined as the fractional change in resistance divided by the strain. For example, 2 microstrain applied to a gage with gage factor of 2 produces a fractional resistance change of $(2x2)10^{-6} = 4x10^{-6}$, or 4 µΩ. Common gage resistance values typically range from 120 to 350 Ω, but some devices can be as low as 30 Ω or as high as 3 kΩ.

STRAIN MEASUREMENT CONFIGURATIONS

Wheatstone Bridge

To make an accurate strain measurement, extremely small resistance changes must be measured. The Wheatstone bridge circuit is widely used to convert the gage's microstrain into a voltage change that can be fed to the

Full-Bridge Circuit

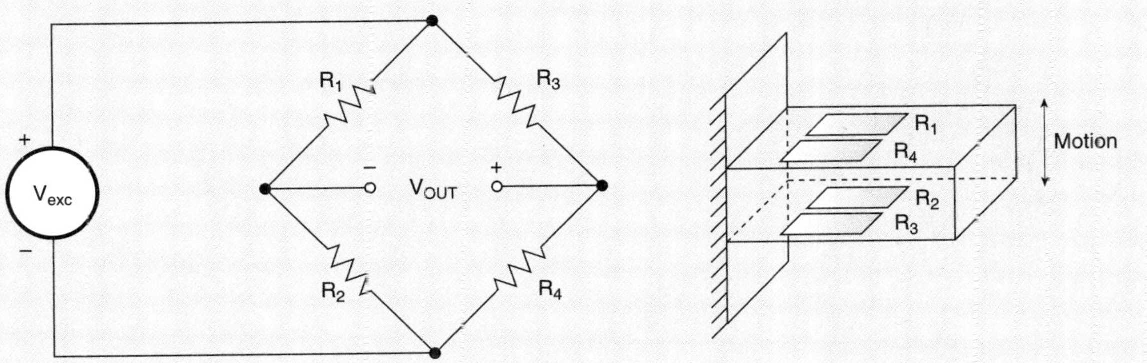

Fig. 7.01. The full-bridge circuit provides the largest output with minimum errors. All four arms of the bridge are active; two are in tension while the two on the opposite side are in compression.

input of the ADC. (See Figure 7.01.) When all four resistors in the bridge are absolutely equal, the bridge is perfectly balanced and $V_{out} = 0$. But when any one or more of the resistors change value by only a fractional amount, the bridge produces a significant, measurable voltage. When used with an instrument, a strain gage replaces one or more of the resistors in the bridge, and as the strain gage undergoes dimensional changes (because it is bonded to a test specimen), it unbalances the bridge and produces an output voltage proportional to the strain.

Full-Bridge Circuits

Although half-bridge and quarter bridge circuits are often used, the full bridge circuit is the optimal configuration for strain gages. It provides the highest sensitivity and the fewest error components, and because the full bridge produces the highest output, noise is a less significant factor in the measurement. For these reasons, the full bridge is recommended when possible.

A full bridge contains four strain gages mounted on a test member. (See Figure 7.01.) Two gages are mounted on the surface under tension and the other two are mounted on the opposite surface under compression. As the member deflects, the two gages in tension increase in resistance while the other two decrease, unbalancing the bridge and producing an output proportional to the displacement. The bridge output voltage is given by:

EQN: 7.01. Full-Bridge Output Voltage

$$V_O = (V_{ex})(X)$$

Where:

V_O = bridge output voltage, V

V_{ex} = excitation voltage applied to the bridge, V

X = relative change in resistance, $\Delta R/R$

The bridge nulls out potential error factors such as temperature changes because all four strain gages have the same temperature coefficient and are located in close proximity on the specimen. The resistance of the lead wire does not affect the accuracy of the measure-

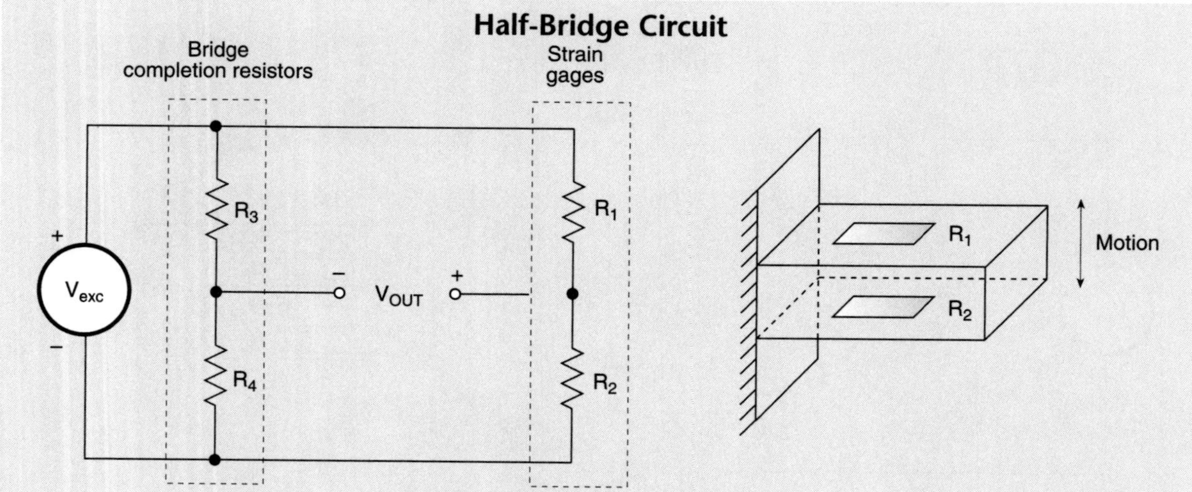

Fig. 7.02. In a half-bridge circuit, only two arms are active. Two strain gages are on the specimen while the two fixed resistors that complete the bridge are not.

ment as long as the input amplifier has high input impedance. For example, an amplifier with a 100-MΩ input impedance produces negligible current flow through the measurement leads, minimizing voltage drops due to lead resistance.

Half-Bridge Circuits

When physical conditions do not allow mounting a full-bridge gage, a half bridge might fit. Typically, two strain gages are mounted on a test member, and two discrete resistors complete the bridge. The output voltage is:

EQN. 7.02. Half-Bridge Output Voltage

$V_O = V_{ex}\,(X/2)$

Where:

V_O = bridge output voltage, V

V_{ex} = excitation voltage applied to the bridge, V

X = relative change in resistance, $\Delta R/R$

For a large ΔR, half bridge and quarter-bridge circuits can introduce an additional nonlinearity error. (See Figure 7.02). Also, the readings are not accurate when the temperature coefficients between bridge completion resistors and strain gages are different and the resistances do not change proportionally with temperature. Furthermore, bridge completion resistors are not usually located near the strain gages, so temperature differences contribute additional errors. In systems with long lead wires, the bridge completion resistors should be attached close to the gages; however, this may not always be practical due to test fixture limitations or other physical conditions.

Quarter-Bridge Circuits

A quarter-bridge circuit uses one strain gage and three bridge completion resistors. The output voltage is:

EQN. 7.03. Quarter-Bridge Output Voltage

$V_O = V_{ex}\,(X/4)$

Where:

V_O = bridge output voltage, V

V_{ex} = excitation voltage applied to the bridge, V

X = relative change in resistance, $\Delta R/R$

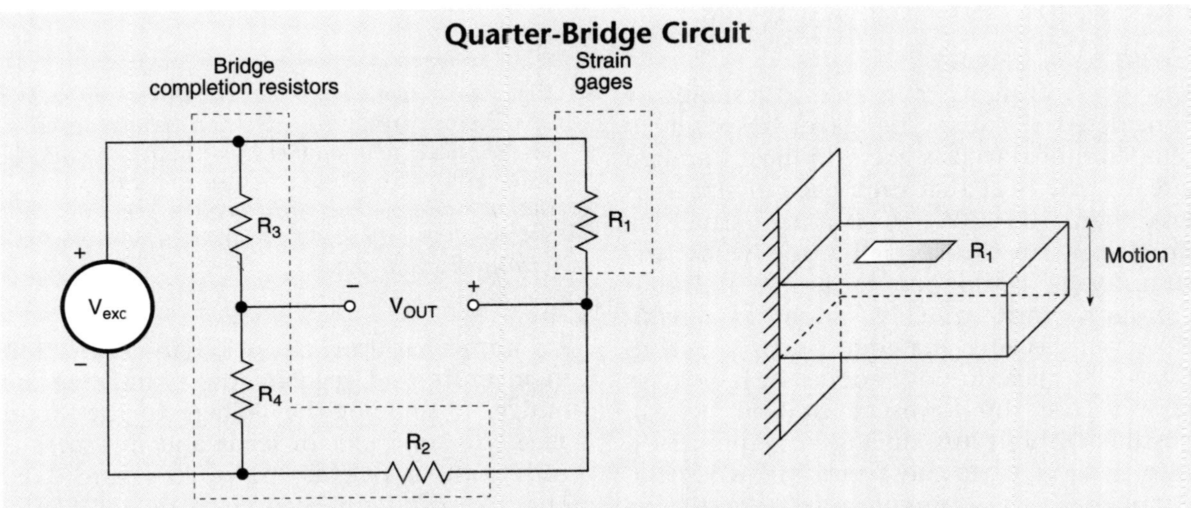

Fig. 7.03. A quarter-bridge circuit uses only one active arm and is the least sensitive of the three types. It is also the most prone to noise and errors.

This arrangement has the smallest output, so noise is a potential problem. Furthermore, all the error sources and limitations in the half-bridge apply to the quarter bridge circuit. (See Figure 7.03.)

Excitation Source

Accurate measurements depend on a stable, regulated, and low-noise excitation source. A regulated source is necessary because the output voltage of a strain gage is also proportional to the excitation voltage. Therefore, fluctuations in the excitation voltage produce inaccurate output voltages.

An ideal data acquisition system provides an excitation source for each channel, independently adjustable from 0.5 to 10.5 V with a current limit of 100 mA. An excitation voltage, V, used with a strain gage of resistance, R, requires a current of I = V/R. The resistance of a Wheatstone bridge measured between any two symmetrical terminals equals the value of one of the resistance arms. For example, four 350-Ω arms make a 350-Ω bridge. The load current equals the excitation voltage divided by the bridge resistance; in this case, 10 V/350 = 0.029 A = 29 mA.

Heating

Resistive heating in strain gages also should be considered because the gages respond to temperature as well as stress. In most standard circuits, the heat that each gage dissipates is less than 100 mW, so it's not usually a problem. This is especially true when the strain gage is bonded to a material that conducts heat quickly, such as metal. However, because materials such as wood, plastic, or glass do not conduct heat away as rapidly, use the lowest excitation voltage possible without introducing noise problems. Also, heat can become a problem when the strain gages are uncommonly small, or numerous gages occupy a limited space.

Consider a Kelvin connection for applying the excitation voltage. Because the excitation leads carry a small current, they drop a correspondingly small voltage; $V = I/R_L$, which reduces the voltage reaching the bridge terminals. As illustrated in Figure 7.04, Kelvin connections eliminate this drop with a pair of leads added at the excitation terminals to measure and regulate the bridge voltage. For example, when i_e = 50 mA, R_L = 5 Ω, and the combined voltage drop in the two leads is 500 mV, no voltage drops in the sense wires.

Kelvin Strain-Gage Bridge Circuit

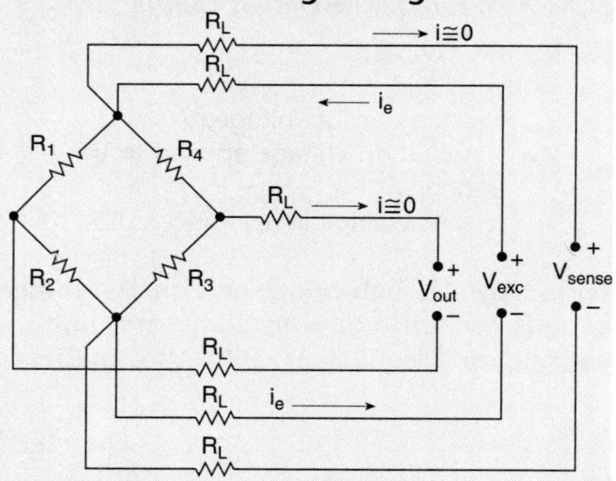

***Fig. 7.04.** The Kelvin bridge circuit uses one pair of wires to provide the excitation voltage directly at the bridge, and another to sense the excitation voltage. A third pair of wires measures the bridge output voltage. This arrangement removes the voltage-drop error in the excitation wires from the measured strain signal.*

A commercial unit uses a Kelvin connection to measure and regulate the voltage at the bridge. It supplies the voltage to the strain gage with one pair of leads and measures it with another pair as shown in Figure 7.05. The six wires are used in pairs for Sense, Excite, and Measure. The Sense lead is a feedback loop to ensure that the Excite voltage is constantly held within specifications.

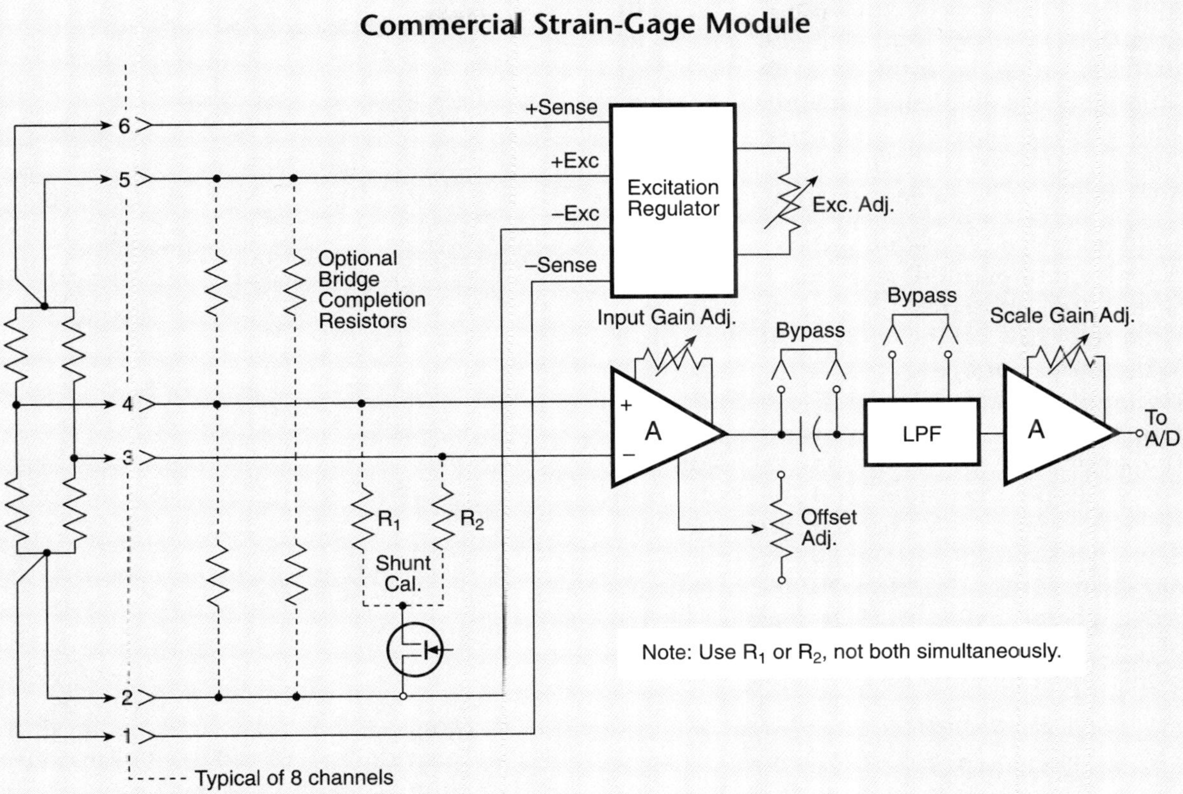

Fig. 7.05. A commercial strain gage module provides adjustable excitation, gain, and offset for each channel. This lets it make use of the instrument's entire dynamic range.

Strain Gage Signal Conditioning

Most strain gage-based transducers and load cells are assigned units of measure for weight, force, tension, pressure, torque, and deflection with a full-scale value measured in mV/V of excitation. For example, a load cell with a 10-V excitation supply and a 2-mV/V-gain factor generates an output of 20 mV at full load, whether the load cell was designed to handle 10, 100, or 1,000 lbs. The difference is in the resolution of the system. That is, the small 10-lb load cell produces 0.5 lbs/mV, and the large 1,000-lb load cell produces 50 lbs/mV.

Conductors carrying such low level signals are susceptible to noise interference and should be shielded. Low-pass filters, differential voltage measurements, and signal averaging are also effective techniques for suppressing noise interference. Furthermore, instrumentation amplifiers usually condition the extremely low strain gage signals before feeding them to ADCs. For example, a 10-V full-scale input provides 156 μV of resolution for a 16-bit ADC. The instrumentation amplifier gain should be adjusted to provide the full-scale output of the strain gage or load cell over the entire range of the ADC.

Force and pressure transducers typically generate an offset output signal when no external force is applied. Instrumentation amplifiers usually contain a control to adjust this offset to zero and let the load cell cover the full range of the ADC. Most instruments also provide adjustable excitation, and gain.

Common Mode Rejection Ratio

A high common mode rejection ratio (CMRR) is essential for strain gage amplifiers. (See Chapter 5.) A strain-gage signal in a Wheatstone bridge is superimposed on a common-mode voltage equal to half the excitation voltage. CMRR is a measure of how well the amplifier rejects common-mode voltages. For example, consider a 10-V excitation supply (V_{max} = 5 V) for a strain gage with 2 mV/V (V_S = 20 mV) at full scale and an amplifier with a CMRR of 90 dB. (See Figure 7.06.) The amplifier can introduce 0.158 mV of error, corresponding to about 0.80% full scale, which may not be acceptable:

EQN: 7.04. Common-Mode Rejection Ratio

$$\text{dB} = 20\log_{10}(V_S/V_e)$$

$$V_{max}/V_e = \log_{10}^{-1}(\text{dB}/20) = \log_{10}^{-1}(90/20) = 31{,}622$$

$$V_e = V_{max}/\log_{10}^{-1}(\text{dB}/20) = 5.00/31{,}622 = 0.158\text{ mV}$$

$$\%\text{ error} = (V_e/V_S)100 = (0.158/20)100 = 0.79\%$$

Where:

V_e = error voltage, 0.158 mV

V_S = signal voltage, 20 mV

V_{max} = maximum voltage, 5V

CMRR = 90 dB

By comparison, a CMRR of 115 dB introduces only 9 µV of error, which corresponds to only 0.04% of full scale.

Wheatstone-Bridge Circuit

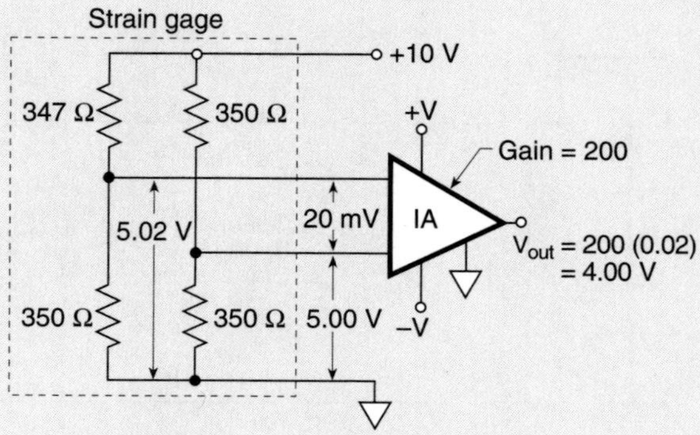

***Fig. 7.06.** Normally the strain gage signal in a Wheatstone bridge is superimposed on a common-mode voltage equal to half of the excitation voltage. Consequently, a high CMRR is necessary to reject the common-mode voltage and amplify the strain gage signal.*

Strain gage signal-conditioning modules usually provide a regulated excitation source with optional Kelvin excitation. Onboard bridge-completion resistors may be connected for quarter and half-bridge strain gages. Instrumentation amplifiers provide input and scaling gain adjustments and an offset adjustment nulls large quiescent loads. This lets input signals use the full range of the data acquisition system and the measurements cover the full resolution of the ADC.

Some strain gage signal conditioners provide fixed gain, offset, and excitation settings, but fixed settings do not take advantage of the maximum dynamic range of the ADC. It decreases the actual available resolution of the measurement. For example, many generic strain gage signal conditioner modules can be set to a fixed 3-mV/V rating. At 10 V, the excitation, offset, and gain trimming are all fixed and no adjustments can be made.

An excitation adjustment lets users set the excitation voltage to the maximum allowed by the manufacturer, which maximizes the bridge's output. Also, the offset adjustment lets users zero the output offset produced by either a small bridge imbalance or a quiescent deformation of the mechanical member. And the gain adjustment lets users set a gain that provides a full-scale output under maximum load, which optimizes the dynamic range of the ADC.

Calibration

The signal-conditioning module also typically provides a shunt calibration feature. (See Figure 7.07.) It lets users switch their own shunt resistors into either one of the two lower legs of the bridge under software control. For example, a shunt resistor can be calculated to simulate a full load. Applying a shunt resistor is a convenient way to simulate an imbalance without having to apply a physical load. For any balanced bridge, a specific resistor can be connected in parallel with one of the four bridge elements to obtain a predictable imbalance and output voltage. For example, a 350-Ω, 2-mV/V strain gage delivers full output when one leg drops by 0.8% to 347.2 Ω. A 43.75 kΩ resistor shunted across one or the other lower bridge elements swings the output to full positive or full negative.

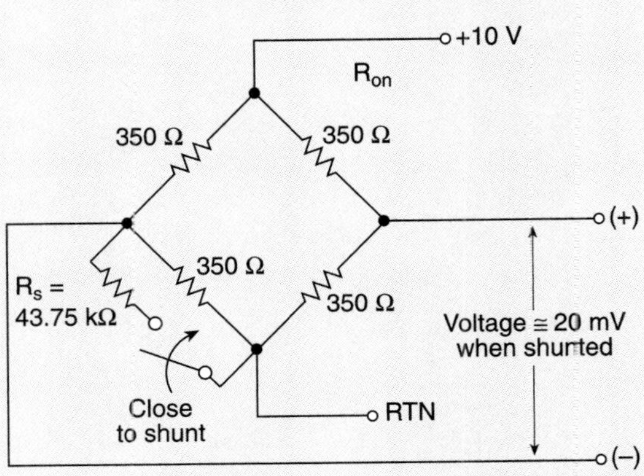

Fig. 7.07. A bridge may be calibrated with a shunt resistor switched with a software command into either one of the two lower legs.

An equation for calculating the shunt calibration resistor value is:

EQN. 7.05. Shunt Calibration Resistor for Transducers

$$R_S = R_{ba}[V_{ex}/4(V_O)]$$

$$R_S = 350[10/4(0.020)] = 43.75\text{ k}$$

Where:

R_S = shunt resistor, Ω

R_{ba} = bridge arm resistor, Ω

$R_S >> R_{ba}$

V_{ex} = excitation voltage, V

V_O = bridge output voltage, V

Many products include calibration software with a Windows-based program that provides several calibration methods, online instruction, and a diagnostic screen for testing the calibrated system.

Transducers and Load Cells

Strain gages are commercially available in prefabricated modules such as load cells that measure force, tension, compression, and torque. Load cells typically use a full-bridge configuration and contain four leads for bridge excitation and measurement. The manufacturers provide calibration and accuracy information.

Strain Diaphragm Pressure Gages

A strained-diaphragm pressure gage consists of two or four strain gages mounted on a thin diaphragm. The gages are wired in a Wheatstone bridge circuit, including bridge completion resistors when needed, so the pressure gage is electrically equivalent to a load cell. The output voltage is specified in mV/V of excitation for a full-scale pressure differential across the diaphragm.

When one side of the diaphragm (called the reference pressure side) is open to the ambient atmosphere, the gage compares the inlet pressure to the ambient pressure, which is about 14.7 psi at sea level. When the gage measures ambient pressure, the reference chamber must be sealed with either a vacuum reference (near zero psi) or the sea-level reference.

Temperature variations can affect the accuracy of the strain diaphragm pressure gages. A pressure gage with a sealed non-zero reference pressure exhibits temperature variations consistent with the ideal gas law. For example, a 5°C change in ambient temperature near normal room temp (25°C) produces an error of 1.7% in the pressure measurement. Temperature variations can also affect the performance of the strain gages themselves. Transducers must contain temperature compensation circuits to maintain accurate pressure measurements in environments with widely varying temperatures.

All strained-diaphragm pressure gages require a regulated excitation source. Some gages contain internal regulators, so users can connect an unregulated voltage from a power supply. Some strained-diaphragm pressure gages also employ internal signal conditioning, which amplifies the mV signal output of the Wheatstone bridge to a full-scale voltage from 5 to 10 V. Gages of this type have low-impedance outputs. In contrast, other pressure gages have no internal signal conditioning so their output impedance equals the Wheatstone bridge resistance (several kΩ for semiconductor types), and their full-scale output is in mV.

Chapter 8

VIBRATION AND SOUND

SENSORS FOR VIBRATION AND SOUND MEASUREMENTS

Strain-Gage Accelerometers

Strain-gage accelerometers contain a small, calibrated mass attached to the free end of a cantilever beam. (Also see chapter 7.) One strain gage sits on the top of the cantilever beam, and another sits opposite it on the underside. Both strain gages measure the bend of the cantilever, one in compression and the other in tension when the device accelerates in a direction perpendicular to the plane of the gages. First, the product of the mass and acceleration converts to force (F = ma) then force converts to an electrical signal. The gages are calibrated to generate an output signal proportional to the amount of acceleration, typically measured in gravity units or gs. The accelerometers' natural frequency is usually in the order of several kHz, and in an undamped system, the cantilever-mass combination forms a harmonic resonator with Q near 100,000.

A system with an extremely high Q excited near its resonant frequency produces large oscillations, which can mask the true acceleration signal, produce inaccurate measurements, and damage the sensor. Consequently, oil or a similar material is frequently added to control the damping. Figures 8.01 and 8.02 show the frequency response of harmonic resonators with Q = 1 and 100,000. For Q = 100,000, the accelerometer is accurate within 10% to 1/3 of its resonant frequency. In contrast, for Q = 1, the accelerometer is accurate within 10% to half of its resonant frequency.

Accelerometer Frequency Response

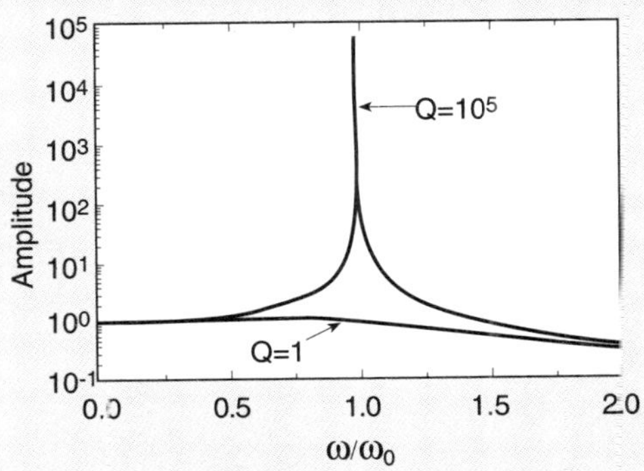

Fig. 8.01. *An undamped accelerometer can have a high Q at its resonant frequency, which produces large oscillations and completely overrides the acceleration signal. Also, if large enough, resonant oscillations can destroy the accelerometer.*

Low Frequency Response

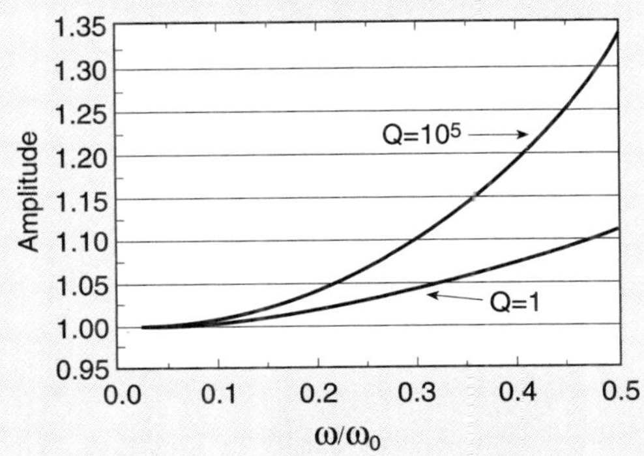

Fig. 8.02. *The low frequency portion of the signal shown in Figure 8.01 is magnified to show its response.*

Most strain gages in accelerometers are wired in Wheatstone bridge circuits. Signal conditioners for a common strain-gage bridge also apply to this type of transducer since it resembles a strain-gage circuit in a load cell.

Piezoelectric Transducers

Piezoelectric materials also are used in accelerometers and microphones. They generate electrical charges on opposite faces of the crystal under dynamic mechanical forces including compression, tension, and twisting. A variety of transducers, such as microphones use piezoelectric elements to convert sound energy picked up by a diaphragm into electrical signals. Conversely, sonic transducers use piezoelectric elements to convert electrical signals into sound energy. Quartz is one of the most common materials applied in piezoelectric transducers and are readily available. Another common material is a piezoceramic material, composed of lead, zirconate, and titinate (PZT).

From an electrical viewpoint, a piezoelectric device resembles a capacitor containing a time varying charge, Q(t). The charge is proportional to the force on the crystal and is usually measured with either a voltage or charge amplifier.

Voltage Amplifier

In Figure 8.03, the voltage amplifier's gain is 1. Adding feedback resistors or increasing the number of amplifier stages changes the gain. The amplifier converts the high-impedance voltage input to a low-impedance voltage output. The voltage is Q/C, where Q is the charge in coulombs, and the C is the capacitance in Farads, which includes both the sensor and the lead capacitance. When calibrating a charge amplifier, the cable capacitance is part of the charge equation, so replacing the cable without recalibration can change the instrument's output accuracy.

Charge Amplifier

The charge amplifier shown in Figure 8.03 is extremely versatile because its output voltage is $V = Q/C$ where Q is the charge, and C is the feedback capacitance. When the operational amplifier has an extremely large open-loop gain, the output voltage is independent of the cable capacitance. This lets the amplifier work well even when it's located several meters away from the sensor. However, because electrical noise increases in proportion to

Piezoelectric Voltage and Charge Amplifier

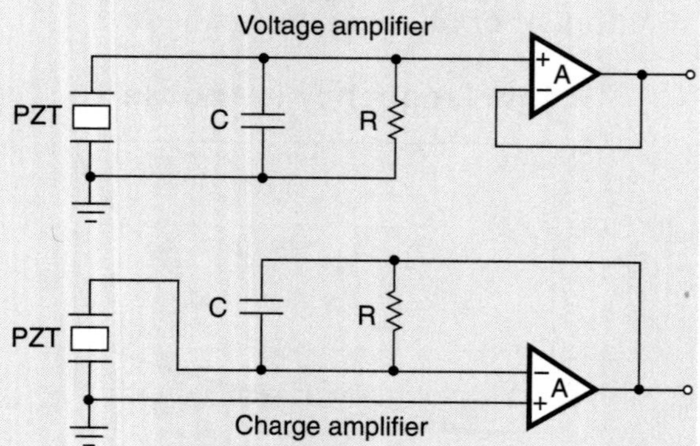

Fig. 8.03. A high-gain op amp isolates the sensor's output voltage from the effect of signal cable capacitance and lets the amplifier reside several meters away from the sensor.

Piezoelectric Sensor and Amplifier

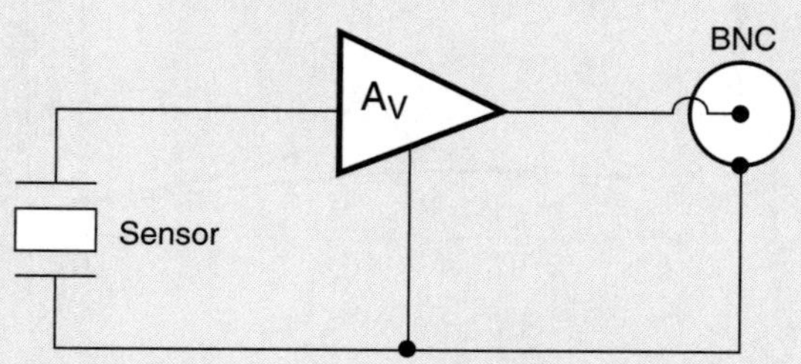

Fig. 8.04. Some piezoelectric accelerometers come with built-in amplifiers. They have a low output impedance and require an external power supply.

cable length, noise susceptibility determines the maximum permissible cable length more than does device sensitivity. The charge amplifier doubles as a high-pass filter with lower corner frequency f = 1/(2πRC), but this represents a trade-off between sensitivity and frequency response. Decreasing C increases sensitivity, but it also increases the lower corner frequency.

PZT-type sensors can't make static measurements because of its leakage resistance and high output impedance, which ranges from 10^{10} to 10^{12} Ω. Such leakage resistance combined with a capacitance of several hundred pF, yields a time constant of a few seconds.

Low-Impedance Transducers

Most modern piezoelectric transducers contain integrated signal conditioning amplifiers. (See Figure 8.04.) Also known as integrated-circuit piezoelectric transducers, these units have a low impedance output and require an external power supply. The manufacturer usually specifies their sensitivity and frequency range. Users need only connect a supply to the power terminals and the output to a voltmeter circuit. The voltage is then scaled to the measured engineering units.

The most common application for piezoelectric-type accelerometers is in measuring a wide range of accelerations and mechanical vibrations. They monitor automobile deceleration (and deploy the air bag at the correct millisecond) in safety systems, lift-off acceleration and motion during space-shuttle missions, and mechanical vibration in numerous machines. Piezoelectric sensors, however, cannot measure constant acceleration as do strain-gage sensors.

Low-impedance piezoelectric transducers also measure pressure or force. The accelerometer circuit requires only two wires to handle both power and signals. Due to the sensor's low impedance, the system is not sensitive to externally introduced or triboelectric-cable noise or cable length. Piezoelectric sensors have resonant frequencies as high as 120 kHz giving them a usable frequency range of less than 1 Hz to more than 40 kHz.

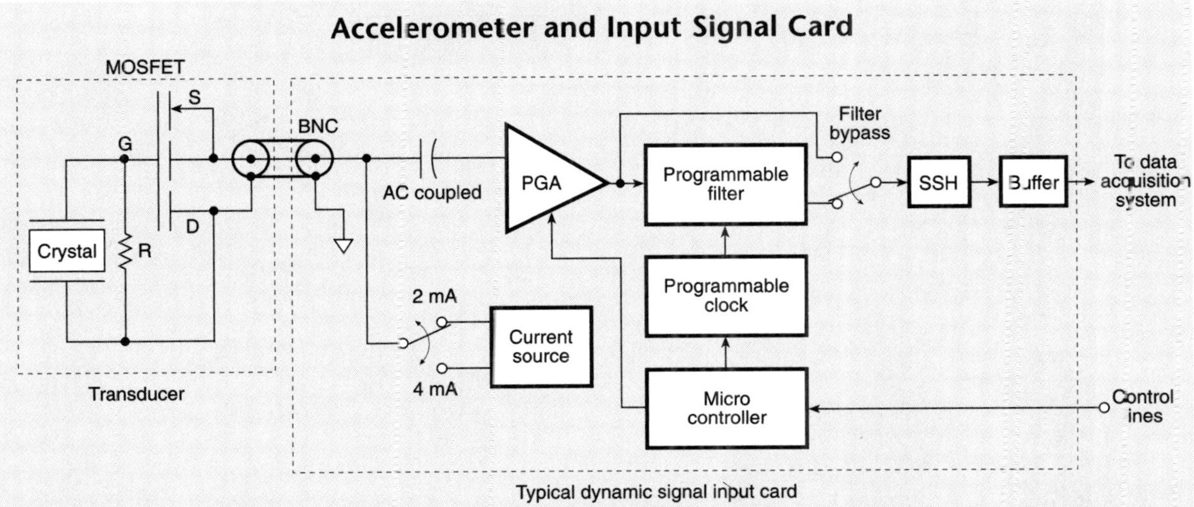

Fig. 8.05. A piezoelectric accelerometer with a built-in MOSFET amplifier stage drives a programmable gain amplifier on a dynamic signal input card. The 2 or 4 mA current source powers the MOSFET.

Figure 8.05 shows a simplified connection scheme between an accelerometer and signal-conditioning card. The voltage developed across R is applied to the gate of the MOSFET, which receives power from a constant-current source of 2 to 4 mA. The MOSFET circuits bias off at approximately 12 V in the quiescent state. When the system is excited, voltage develops across the crystal, which is applied to the gate of the MOSFET. The voltage produces linear variations in the MOSFET's impedance, which, in turn, produces a proportional change in the bias voltage. This voltage change couples to the input amplifier through capacitor C. The value of R and the internal capacitance of the piezoelectric crystal control the low-frequency corner. Units weighing only a few hundred grams generate high-level outputs to 10 mV/g with response to frequencies from 0.3 Hz to 2 kHz. Smaller units with less sensitivity respond to frequencies from 1 Hz to 35 kHz.

The constant-current source provides a source-to-gate bias for the FET. As the gate current responds to changes in applied pressure on the crystal, the drain-to-source voltage (Vds), and the voltage out of the preamplifier change proportionally. An ac coupling circuit or a high-pass filter is always necessary because of the high dc offset at Vds that develops from the bias current. The cutoff frequency of the high-pass filter depends on the application and the particular accelerometer. (See Figure 8.06.)

To eliminate the need for an outboard preamplifier, some accelerometers contain the current source and the ac coupling circuitry. This feature also lets accelerometers connect to the data acquisition system through simple BNC connectors. Most accelerometers, however, require an amplifier and filter on the output before the analog to digital conversion stage. Also, the

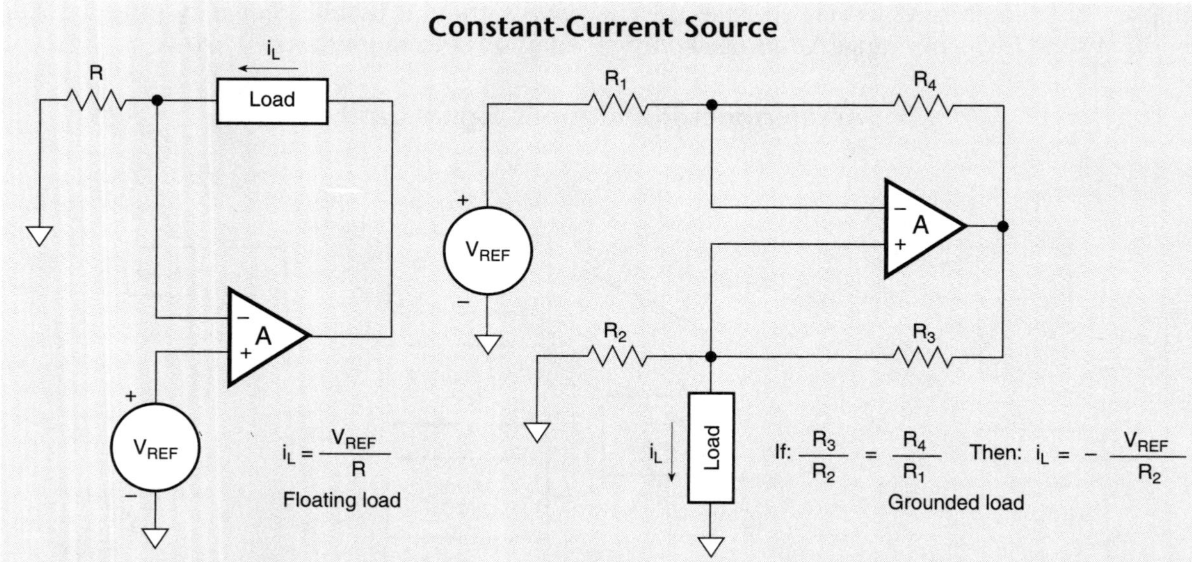

Fig. 8.06. A constant current source provides a source-gate bias for the field-effect transistor in the accelerometer. Depending on the specific FET circuit, a constant-current source may be configured for either a floating load or a grounded load.

programmable-gain amplifier lets the operator adjust the gain for optimum response.

Along with programmable amplifiers, programmable low-pass filters reject unwanted high-frequency signals. The signals typically come from noise or high-frequency vibrations that do not relate to the application. When developing the front-end circuitry for this type of measurement, noise rejection and bandwidth are primary concerns. As the bandwidth increases, the noise can increase as well. Low-pass filters also should be used in most accelerometer conditioning circuits to reduce noise and aliasing effects. The cutoff frequency of the low-pass filter should be close to the system's maximum useful operating frequency.

Chapter 9

DISPLACEMENT AND POSITION SENSING

LINEAR DISPLACEMENT

Motion control systems typically require a sensor in a feedback loop to ensure that the moving member or component reaches its commanded position. Position sensors are typically used on machine-tool controls, elevators, liquid-level assemblies, forklift trucks, automobile throttle controls, and numerous other applications. Position sensors measure either absolute or incremental displacement, and are made with a variety of different materials. Also, when the sensor's power fails, some retain the measured position data, and other types lose the information.

One common device is a resistive position sensor or potentiometer. It may be a single-turn or a multi-turn rotary sensor with an element made of carbon or any one of a number of more durable, high-temperature, conductive-plastic film materials. Some types of potentiometers connect to the moving member through a rigid or flexible coupling while others are actuated with a string wrapped around its output shaft. Yet other position sensors include LVDT's (linear variable differential transformers), sonically operated devices, and digital encoders.

String-Potentiometer Transducers

String-type potentiometers are basically rotational devices that typically measure linear motion as shown in Figure 9.01. Because a cable connects between the moving member and the potentiometer, the cable must be preloaded with a spring, have a low coeffi-

String-Potentiometer Position Sensor

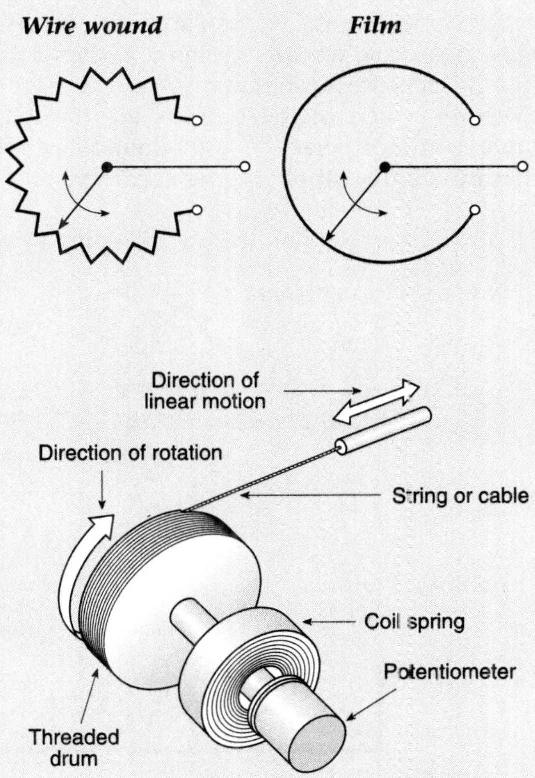

Fig. 9.01. String potentiometers occupy relatively little overall space in a machine installation or device under test. In some applications where a sensor cannot be mounted on the device being measured, the potentiometer may be conveniently located some distance from the test bed without sacrificing accuracy or repeatability.

cient of expansion under wide ranges of temperature (–65 to +125°C), and have little tendency to stretch. Because of these constraints, its accuracy is somewhat less than that of a more rigid, shaft-coupled sensor.

Most modern potentiometers are extremely linear and are made with special alloys or precious metal wipers and conductive plastic resistive elements to withstand as many as 100 million revolutions. They are usually part of a voltage divider and provide an output voltage proportional to displacement, but they may also drive analog-to-digital converters to feed microprocessor-based instruments directly. However, when digital outputs are needed, a more common practice uses digital encoders that inherently supply higher accuracy.

LDVT: Linear Variable Differential Transformers

LVDTs are electronic devices comprised of a special transformer with a movable metal actuator in its hollow-core. The transformer contains two differentially wound secondary windings on either side of a primary winding. As the actuator in the core moves in and out, it couples a signal from primary to secondary in proportion to the displacement. Being differentially wound, LVDTs also can provide directional signals with the appropriate signal conditioning circuits. With the actuator at the center position, the output signal is null or zero. As the actuator moves out in one direction, the signal is positive, and when it moves in the opposite direction from null, the signal is negative. (See Figure 9.02.)

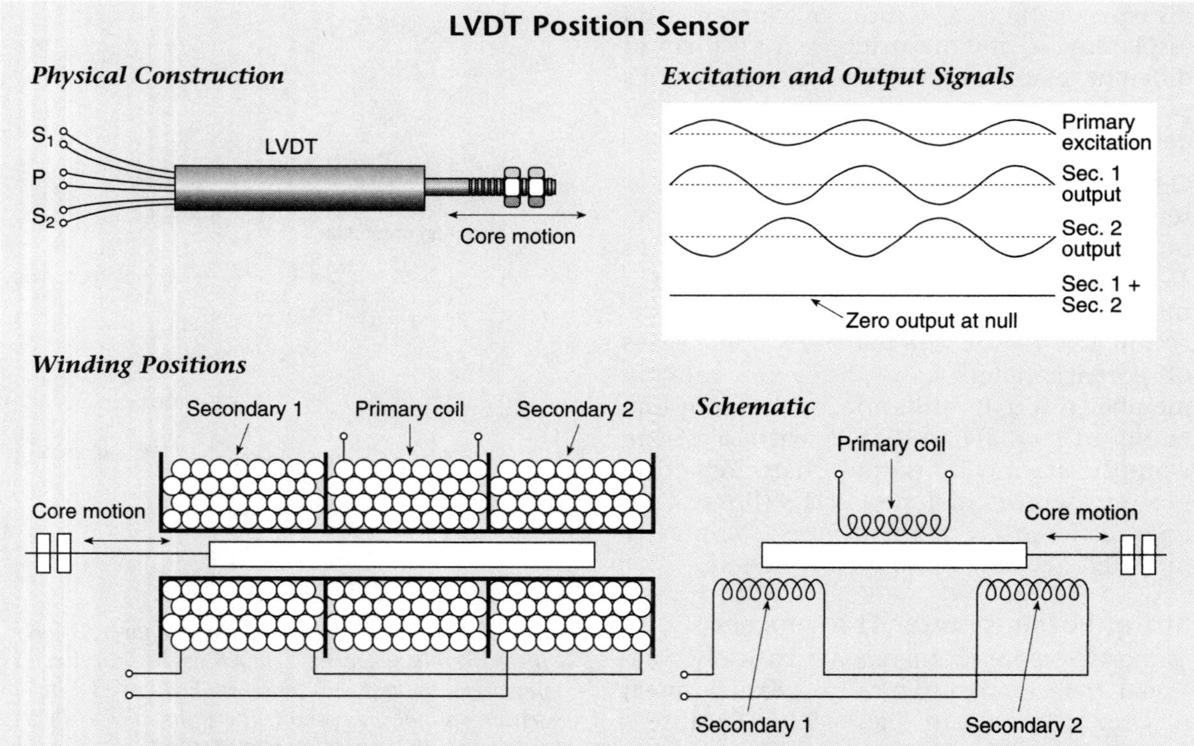

Fig. 9.02. LVDT sensors provide distance and direction information with extremely high accuracy. But a drawback for some installations is their relatively long length. The minimum length of the sensor is its body, and the core can extend only as far as the length of the body.

LVDTs require electronic circuits to provide an ac signal to excite the primary and a demodulator to convert the output to a dc signal with directional polarity. Some LVDTs contain the electronics onboard and only require power to obtain an output, while others require separate signal conditioners or modules. The LVDTs themselves are rugged devices and have a MTBF of as much as 3 million hours. The main limitation is the displacement range of the core to ensure high linearity.

Ultrasonic Sensors

Another relatively new type of position sensor is based on the theory of operation of the interaction of two magnetic fields on magnetostrictive materials. The major components of the sensor include an external permanent magnet, integral sensing and output signal conditioning circuits, a sensor tube, and a wave-guide element. The permanent magnet fastens to the movable object, and the housing remains attached to a stationary reference point. (See Figure 9.03.)

The output signal conditioning circuit generates a current pulse and magnetic field which travel down the waveguide. When the transmitted field reaches the permanent magnet field, the combination induces a strain pulse in the magnetostrictive tube. The strain pulse, in turn, travels down the waveguide to the head of the sensor where it is detected. The time measured between the application of the interrogation pulse and the arrival of the strain pulse determines the precise, absolute distance between the movable magnet and the pulse generator. Resolution

Magnetostrictive Position Sensor

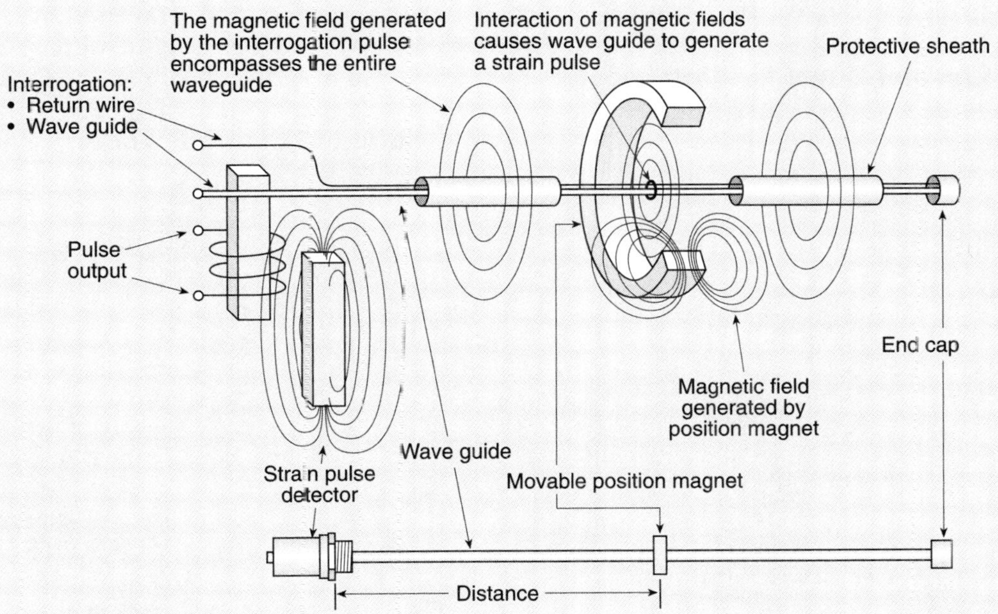

Fig. 9.03. Magnetostrictive position sensors have a high signal-to-noise ratio, shock and vibration resistance, and accuracy. They are rugged devices that deliver from 30 to 400 mV signals at their output terminals and require little or no additional signal conditioning. A possible disadvantage for some applications, however, is its wave-guide, which must be at least as long as the measured stroke.

is about 0.002 mm, durability is high, and the stroke ranges from 5 to 20,000 mm. Higher resolutions require additional interpolating circuits. The output signal, which is proportional to displacement, can be either analog or digital, and it can represent both position and velocity.

Linear Encoders

Linear encoders have evolved over the years to provide extremely accurate position information. Their principle of operation depends upon sensing position from precision graduations (position codes) etched on a linear scale. They come in both contacting and non-contacting types, and their length is primarily limited to their temperature-sensitive coefficient of expansion to preserve accuracy. Most modern scales are glass, and optical sensors detect the graduations. (See Figure 9.04.) The benefit of a non-contacting type is typically longer life because it uses optical components instead of brushes or a contacting wiper. All optical encoders contain a light source, light detector, code wheel or scale, and signal processor.

Two-Channel Linear Position Encoder

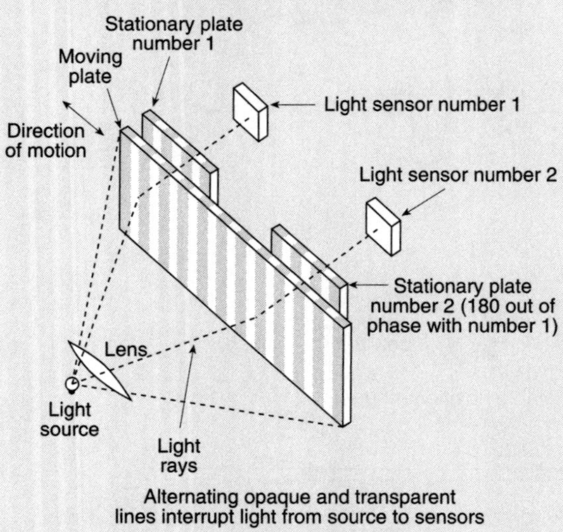

Fig. 9.04. Two-channel encoders provide step and direction vectors with an output signal that can be a sinewave, squarewave, or a series of equally spaced pulses generated at regular intervals on the waveform.

ROTATIONAL POSITION

Rotational position may be measured with potentiometers, coupled to a rack and pinion assembly or gear trains to obtain a suitable ratio of potentiometer revolutions to measured component rotation or movement (see figure 9.05). Potentiometers commonly come in single turn to as many as 25 turns and fit numerous applications. But, most rotational measurements, whether they are velocity, rpm, or position, are best handled with digital encoders. Like linear encoders, several types have been used over the years including both contacting (see Figure 9.06) and non-contacting. They contain a disc (code wheel) or a plate made with alternating opaque and transparent segments that move between a light source such as an LED, and a photo-detector. (See Figure 9.07.)

Rack and Pinion Encoder Assembly

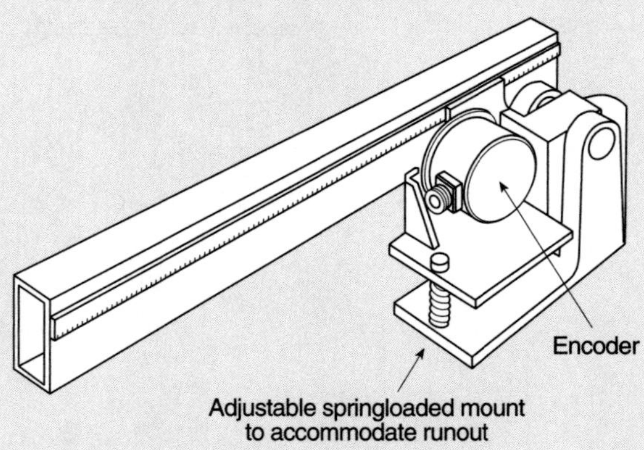

Fig. 9.05. A rack and pinion sensor assembly should have a flexible mount to compensate the combined runout error. In a machine application, the sensor should be protected from metal chips and other foreign objects to ensure measurement accuracy.

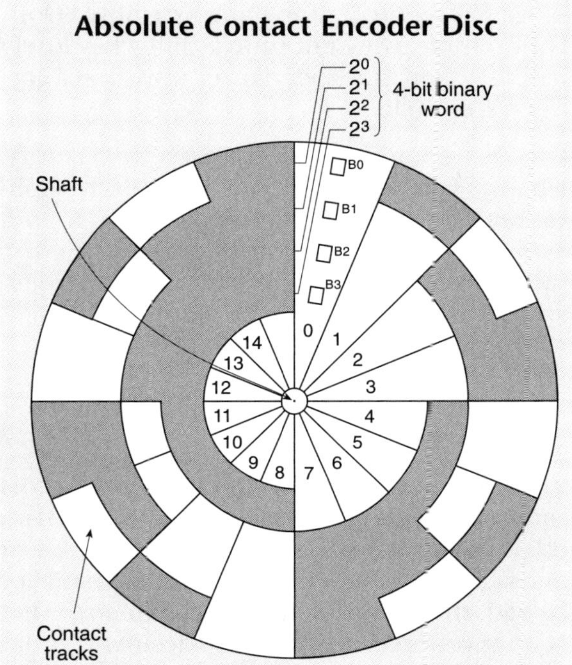

Fig. 9.06. The disc pattern on the contacting encoder is a digital word, which represents the absolute position of the encoder shaft.

Absolute Encoders

Absolute encoders typically contain several detectors and slots on a revolving wheel to provide a unique output binary code for each shaft position within its range of resolution. Therefore, the shaft position can be determined absolutely. Moreover, the correct position is always retained before and after a power failure. The tracks on the encoder wheel may be arranged to provide a 0.5 bit maximum error. Absolute encoders come in single and multi-turn versions.

Incremental Encoders

Incremental encoders usually provide two channels of square wave output signals. Each pulse represents an increment of rotation of relative position. These encoders often contain a third slot called the reference slot used to calibrate the encoder. Dual channel encoders are more accurate than single channel types. They register direction, and when a code wheel stops on the edge of a slot, incident vibrations cannot generate an error count, as it does for single-channel sensors.

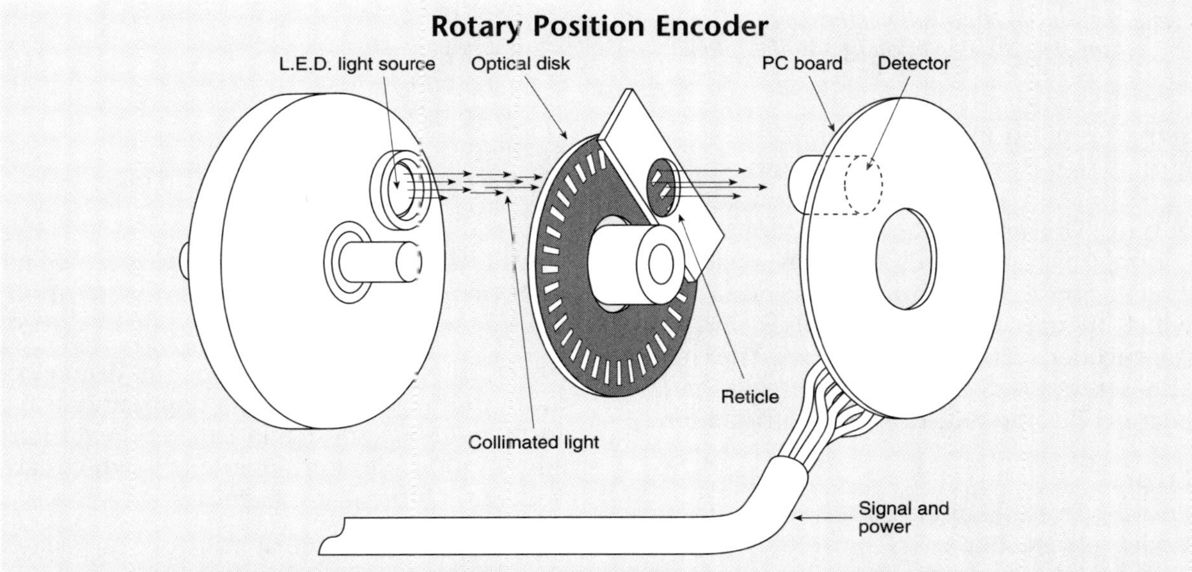

Fig. 9.07. The shutter is one of the most critical components of the encoder assembly. The sensor's resolution and accuracy hinge on the quality of the vendor's manufacturing and assembly processes.

Quadrature Encoders

The two channels, A and B, of a dual output incremental encoder are usually displaced by 90 degrees from each other. Therefore, the spacing between pulses is one-half the line width. When the disc contains 1024 lines and 1024 spaces, the sensor generates 4 X 1024 = 4096 pulses per revolution; hence the output signals are delivered in quadrature. (See Figure 9.08.) The angular movement in this arrangement resolves to 5.27 minutes of arc.

Two-Channel Quadrature Output

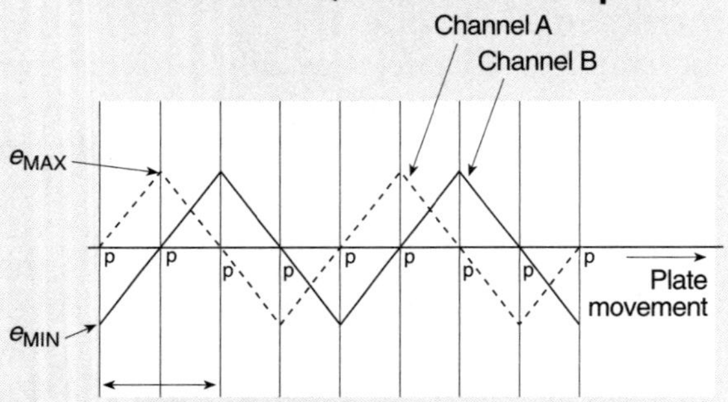

Fig. 9.08. The two light sensors in a dual channel linear position encoder are connected in a push-pull configuration and track one another, so equal changes in the input produce equal changes in the output.

Signal conditioning circuits generate the output signal waveform and complementary signals, sense their direction, and filter the output. The amplitude of the input voltage controls two stable output states of a shaper circuit. The shaper output feeds several gates, which in turn generate the pulses that mark the encoder's zero-signal crossings. The circuits also sense whether channel A leads or lags channel B to provide direction information.

Errors

Encoder manufacturers endeavor to minimize device errors, but users must be concerned with total system errors. Encoder errors consist of instrument and quantization errors, as well as manufacturing and assembly errors. For example, the light sources must be closely matched and the code wheel must have low disc eccentricity.

Encoder error is relatively easy to quantify, whereas total error depends upon each application and is a little more difficult to pin down. Several sources contribute to total error, including manufacturing and assembly tolerances, LED characteristics, scale and disc alignment, and power supply stability. Others include electrical interference, temperature variations, mechanical coupling, and mechanical vibration.

In some cases, encoder misalignment errors can be calculated. For example, a linear sensor often incorporates a glass scale to sense linear displacement. When the scale is bent or bowed, the curvature introduces an error that is proportional to the displacement d, as shown in Figure 9.09. When d is small relative to S, the concave surface of the scale is shortened by Δ, and the convex surface is elongated by Δ, where $\Delta = 4td/S$.

For example, when:

S = 8 in.
t = 0.20 in.
d = 0.0010 in,
then:
$\Delta = (4 \times 0.20 \times 0.0010)/8 = 0.0001$ in.

Lines etched on the convex side produce an error of +0.0001 in., and lines on the opposite side produce an error of –0.0001 in./foot.

Power supply stability and freedom from electrical interference are critical requirements for encoders. Both anomalies can produce intermittent and random operation, which are frequently difficult to diagnose.

Temperature has minimal influence on today's encoder electronics, although the maximum operating temperature at rated power for LEDs

Misalignment Error in Glass Scales

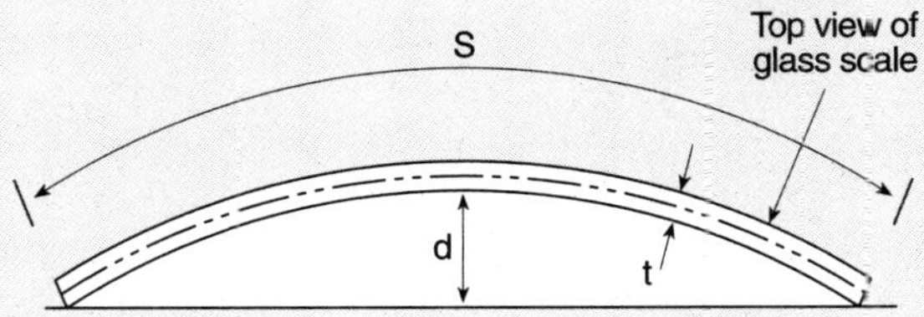

Fig. 9.09. The top view of an exaggerated bow in linear scale illustrates the errors that could undermine measurement accuracy. The error in this case is proportional to the displacement "d."

is 150°F compared to 200°F for incandescent lamps. But LEDs have more than double the expected life of lamps, and they work with superior signal conditioning circuits, so they are used almost exclusively. Temperature fluctuations affect linear glass scales more significantly. For example, a ±150°F temperature change can produce an error of ±0.0008 in./ft.

The mechanical coupling to the moving member should suit the application, considering axial and radial loads, acceleration, velocity, vibration, accuracy, and resolution. The proper coupling reduces errors and increases the encoder's life.

Chapter 10

NOISE REDUCTION AND ISOLATION

CONTROLLING NOISE

Controlling noise in measurement systems is vital because it can become a serious problem even in the best instruments and data acquisition hardware. Most laboratories and industrial environments contain abundant electrical-noise sources, including ac power lines, heavy machinery, radio and TV stations, and a variety of electronic equipment. Radio stations generate high-frequency noise, while computers and other electronic equipment generate noise in all frequency ranges. Building a completely noise-free environment just for running tests and measurements is seldom a practical solution. Fortunately, simple devices and techniques such as using proper grounding methods, shielded and twisted wires, signal averaging methods, filters, and differential input voltage amplifiers can control the noise in most measurements. Some techniques prevent noise from entering the system, while others remove extraneous noise from the signal.

THE GROUNDING CONFLICT

A non-technical dictionary defines the term ground as a place in contact with the earth, a common return in an electrical circuit, and an arbitrary point of zero voltage potential. Grounding, or connecting some part of an electrical circuit to ground ensures safety for personnel and it usually improves circuit operation. Unfortunately, a safe environment and a robust ground system often do not happen simultaneously. It takes planning based on systematically understanding how electricity behaves in various types of circuits. For example, high redundancy is one key feature that makes most of the electrical distribution systems around the world safe and operate properly.

Grounding for Safety

Isolated secondaries of step-down power distribution transformers are generally grounded near the transformer and within the first switching panel in the wired path to the eventual load. (See Figure 10.01.) The ground is a point within the panel connected to a nearby earth ground rod. Typically, a large or significant structure (building frame)

Control Panel Grounding

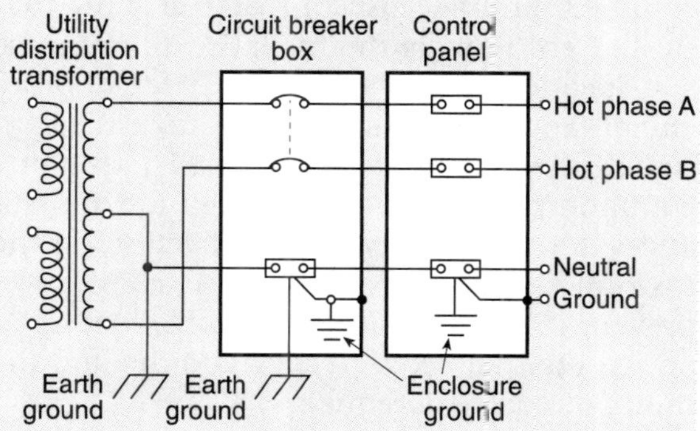

Fig. 10.01. Utility power transformers are typically grounded to earth ground near the transformer, and again at the input to the electrical junction box or first switching panel. The switching panel, in turn, may be connected to a rod driven into the earth to ensure that it too is at true ground potential.

or metallic system (plumbing) is also connected to the same point. This minimizes voltage differences that may develop between a water pipe and an appliance with a three-wire grounded cord, for example. An electrical fault such as a non-grounded conductor contacting a metal object is designed to open a fuse or trip a breaker rather than leave an electrically energized appliance at a higher potential than a nearby water pipe or a sink faucet. If the ground connection in the panel disconnects for any reason, the redundant ground near the transformer will provide the path for fault currents to open fuses or trip breakers. Preventing electrical shocks and electrical fires is the highest priority for ground circuits, but the redundancies built into many electrical grounding systems occasionally limit certain kinds of connections for input to data acquisition systems.

Grounding for Robust Instrumentation

Several internal, common busses in a data control instrument are arranged to regulate current flows and terminate all paths at one common point. This approach ensures that the current flowing in any path will not force a voltage drop in a return path for another circuit and appear as an (erroneous) input signal. (See Figure 10.02.) Usually, this one common point connects through low impedances to the safety ground connection on the instrument's ac power cord. This connection prevents the internal system from floating at an ac potential between earth ground and the input ac supply potential.

Common Grounds

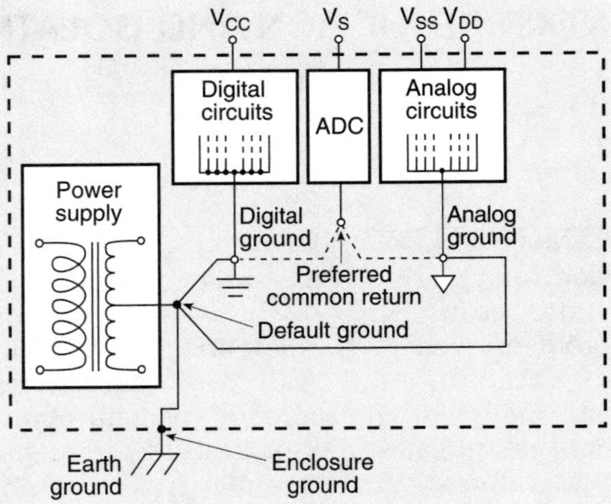

Fig. 10.02. Significant current fed to multiple circuits should have individual return paths to a common ground or negative terminal. This reduces the risk of voltage drops developing over long ground runs or lead wires that could become input (error) signals to other circuits.

GROUND LOOPS

Measuring instruments that contain an earth ground as described above usually generate a ground loop. A ground loop can become a serious problem even when the ground voltage on the measured point equals the ground voltage entering the instrument through the line cord. A voltage that develops between the two grounds can be either an ac or a dc voltage of any value and frequency, and as the voltage and frequency increase, the effects of the ground loop become more troublesome.

Dangerous and Destructive Ground Loops

A transient current can generate a substantial voltage on grounded conductors. During an electrical fault when an energized conductor contacts a safety ground, for example, a fraction of the supply voltage can end up on the safety ground before the fuse or circuit breaker supplying the fault opens and removes the voltage. This happens in a few milliseconds and is usually not a safety hazard. But the problem can be much more serious if lightning strikes a safety ground structure and thousands of amperes flow through the ground system. Potential differences across even a fraction of an ohm can easily exceed 1,000 Vac and damage equipment and endanger lives.

Symptoms of Ground Loops

Sometimes, a measurement error is mistakenly attributed to a ground loop problem, especially where a ground is not strictly involved. The phenomenon relates to two types of situations; shared current flow in a circuit path, which produces unintended voltages, and inadvertent circuits that interfere with the proper operation of intended circuits.

How Ground Loops are Created

A ground loop problem can be illustrated by the following example. An integrated sensor with internal signal conditioning contains three wires; a positive power supply lead, a signal output lead, and a negative lead that serves as both the power return and signal common. (See Figure 10.03.) The sensor's internal circuitry draws about 30 mA and the output signal ranges from 0 to 5 Vdc.

The sensor is stimulated and a digital voltmeter reads the correct output of 2.50 Vdc on the test bench. But when the three leads are extended by 500 feet with 20 AWG wire (10.4 Ω/1,000 ft at 20°C), the common lead wire carrying the 30 mA of power supply current drops about 150 mV. This lead resistance voltage drop adds to the sensors output voltage and delivers 2.65 Vdc to the digital voltmeter. The error amounts to about 6.6% and what's worse, it varies widely with the temperature of the wire. The specific application determines whether the error can be tolerated or not.

When the current drawn by the sensor circuit is not a steady state value, that is, it consists of an ambient level combined with a dynamic component, the error introduced will be time varying. It might be a relatively high frequency, which acts as noise in the measured output, or it may be in perfect synch with the physical phenomenon being sensed. It then affects the magnitude of the time-varying output signal. Both types of errors often appear in data acquisition systems.

How to Eliminate Ground Loops

A reliable trouble-shooting method analyzes current flow and predicts its results. The wires from the intended point of measurement must carry only current associated with the bias requirements of the analog input channel. (See Figure 10.04.) These currents are typically measured in microamperes. At lower voltage levels, they can be altered substantially if forced to share extremely long wires carrying merely mA. A detailed wiring diagram and a circuit schematic can provide insight and understanding to help prevent this type of problem before hundreds of feet of wire are installed.

Interfering Ground loop

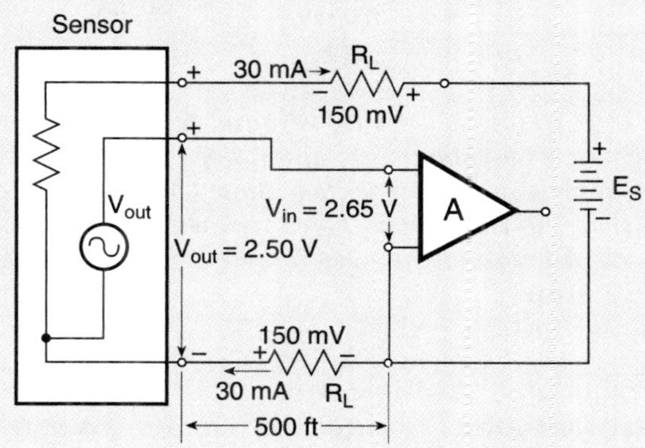

Fig. 10.03. The 150 mV dropped across R_L in the bottom ground return line arises from the 30 mA of current flowing in 5.2 Ω of lead wire resistance. The voltage adds to the sensor's 2.50-V signal to yield 2.65 V at the input to the signal-conditioning amplifier and produces a 6.6% error.

Bypassed Ground Loop

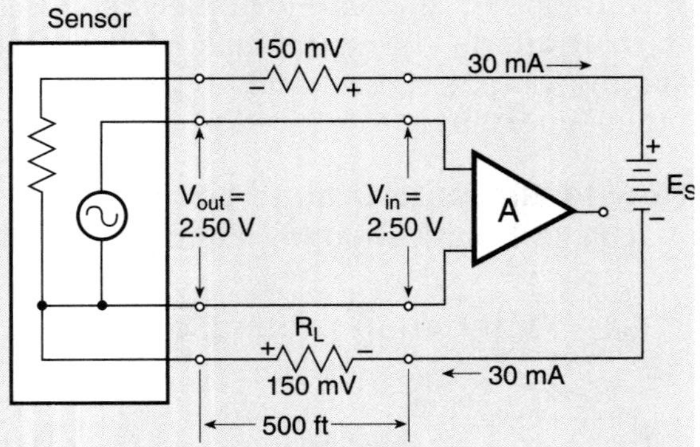

Fig. 10.04. *A separate wire run from the sensor ground (or common terminal) bypasses the power supply ground wire voltage drop. The true output signal of the sensor (Vout) reaches the amplifier input terminals because the input draws negligible current.*

Frequently, multiple wires running between two locations cannot be shared. When a common wire is shared, the current in one channel affects the voltage reading in another channel. In the previous numerical example, a fourth wire connected to the lower end of a differential measurement channel provides an output voltage that can be measured accurately with a high degree of confidence. This approach is most effective when the system supports three wires, and they share a common power supply. Differential input connections used with the analog common, which is referenced to the power supply return terminal, eliminates the effect of the ground loops inherent in this multiple-sensor arrangement.

CROSSTALK IN DATA ACQUISITION SYSTEMS

Another type of ground-loop error is crosstalk between channels. This may be defined as an interaction between readings on two or more channels, which may be static or dynamic. When multiple channels are used and ground loops exist, the simplified errors described previously most likely will be compounded by contributions from other channels. The crosstalk may or may not be obvious.

Static Crosstalk

Consider a group of static channels with steady voltages, that when measured individually, yield accurate readings. However, when each channel is connected to an input of the data acquisition system and the readings change, the change indicates that crosstalk is generated by a steady-state ground loop. Likewise, when the reading of a channel changes by connecting another channel, crosstalk exists and the problem is a ground loop.

Dynamic Crosstalk

Dynamic crosstalk is the name given to the situation where a known dynamic signal on a particular channel appears in a physically unrelated channel. The steady-state currents drawn by the transducers discussed in the previous example are idealized for simplicity. These currents commonly vary with the measured physical variables along with the errors.

Sequentially reading signals of widely varying magnitudes produce sequential crosstalk in multiplexed data acquisition systems. Capacitive or inductive coupling between channels generates crosstalk in systems with improperly or carelessly dressed wires. Generally, however, these are not attributed to ground loops and are less common.

SHIELDED WIRING

Benefits

Metallic shields placed around equipment and test leads effectively prevent noise from either entering or leaving the system. For example, loose or exposed wires become antennas for radio frequency signal pickup, and can form loops that radiate noise.

To emphasize the need for controlling noise, Figure 10.05 shows a single-ended voltage measurement on a shorted channel. Approximately 6 feet of wire, not twisted or shielded, was attached to the data acquisition system. Figure 10.06 shows the noise in a single-ended, shorted channel using shielded cable with obvious improvement.

The best instrumentation wiring schemes consist of carefully grouped lines, twisted in pairs, occasionally covered with a second shield, and routed through a dedicated conduit or raceway. A shielded, twisted pair is quite commonly used in a channel to connect a signal from a source to an input terminal. Shields minimize capacitive coupling and twisted wires minimize inductive coupling.

Proximity to other wires, especially power wires carrying high voltages and high currents can couple noise into low-level signal conductors. Capacitive coupling can exist between any two pieces of metal in close proximity, including two conductors in totally separate circuits. Likewise, air-core transformer coupling can crop up between two closed wiring loops in totally separate circuits.

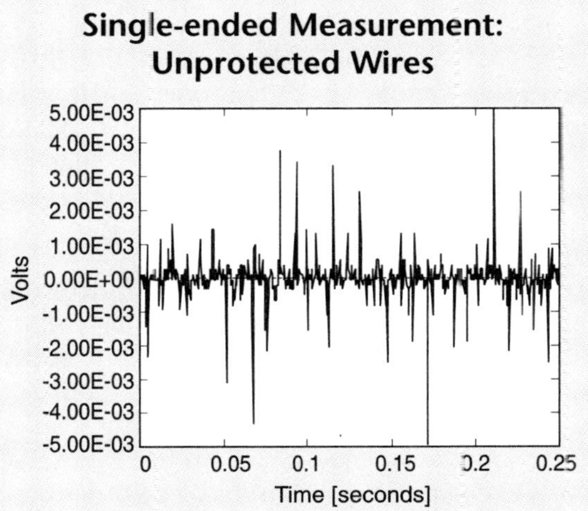

Fig. 10.05. Radiated noise spikes are easily picked up on non-twisted, non-parallel, unshielded lead wires connected to the single-ended amplifier input terminals, even when the inputs are shorted together.

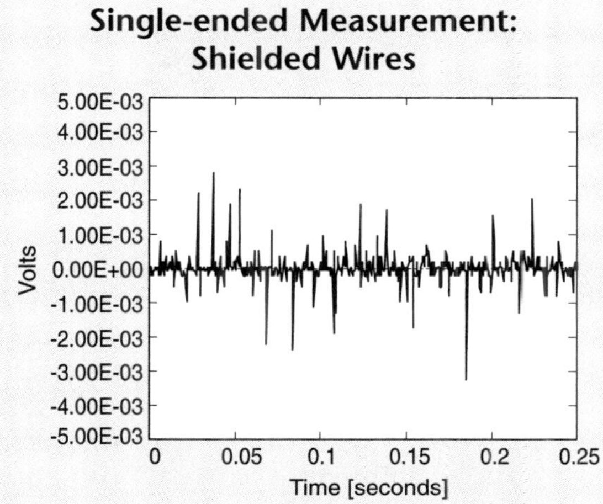

Fig. 10.06. Shielded wires reduce electrical noise pick up at the input of single-ended amplifier input terminals, but other techniques often are more effective.

Proper Installation and Use of Shields

Typically, a shield terminates at one end only, unless it extends to the shield in another span of the same channel wiring. The shield can terminate at either the transducer end or the input channel end, but not both. When the sensor or transducer is in a shielded metallic enclosure, which is also connected to earth ground, the shield may be connected at the sensor end and remain open at the input channel terminals. When the sensor is well insulated, the shield may float and connect to the analog common of the data acquisition system input terminals. Occasionally, multiple-conductor cables composed of a bundle of wires and an overall shield are acceptable for a group of high-level, dc or low-frequency signals, but would not be recommended for the general data acquisition case. Compromising a well planned wiring system with low quality wire, shared conductors or shields, and parallel, untwisted wires will produce less than optimum results.

ISOLATION AND FLOATING DATA ACQUISITION SYSTEMS

Isolation

Isolation is defined as the separation of one signal from another to prevent unintentional interaction between them. All multiplexed data acquisition systems contain a certain degree of channel-to-channel isolation; relay-based systems have galvanic isolation while solid-state systems do not. Galvanic isolation is the absence of any dc path. Most isolation methods eliminate all dc paths below 100 MΩ. Three major benefits of galvanic isolation are circuit protection, noise reduction, and high common-mode voltage rejection, especially those developed by ground loops.

Computer-based data acquisition equipment makes possible an array of multiple channel measurements previously beyond the economic reach of many applications. This has been accomplished by user acceptance of two major compromises, multiplexing and non-isolated inputs. Multiplexing is successful when the sampling rate is adequately high and the source impedances are sufficiently low. Lack of isolation places an entirely different kind of limitation on the type of input signals that can be connected.

Circuit Protection

Isolation separates the signal source from the measurement circuitry that could be damaged by the signal. Voltages higher than about 10 V can distort data or damage components used in the system. High-voltage input signals or signals containing high-voltage spikes should therefore be isolated. The protection also works in the opposite direction to safeguard a sensitive signal conditioner from a device failing elsewhere in the system.

Computer-based data acquisition equipment is most often connected to a host computer, which is connected to earth ground. The analog inputs of plug-in cards and most economical external systems are not electrically isolated from earth ground or each other. Many applications are compatible with this situation, but some applications face a problem with high common-mode voltage.

Rejection of high common-mode voltage

Common-mode input voltage is defined as the voltage applied between the common terminal and the two input terminals with the condition that the two input voltages be identical. In other words, the two input terminals may be connected together and the common-mode voltage applied between the shorted inputs and the common terminal as shown in Figure 10.07. In a practical test and measurement situation, the common-mode voltage may exceed the instrument amplifier's input rating, which is typically less than 10 V. For safe and accurate measurements, common-mode voltages higher than 10 V must be isolated from the instrumentation amplifier while allowing the measured signal to pass. Common types of isolation amplifiers use magnetic, optical, or capacitive means to couple the signal.

Magnetic Isolation

Special instrument amplifiers use transformers that magnetically couple analog-type ac signals from the input section to the output section while effectively sustaining high common-mode voltages. Transformer coupling also lets them provide isolated power to the input stage without using a separate dc/dc converter. A particular instrument amplifier contains an input op amp with a CMRR of about 130 dB at a gain of 100, and 2000 V peak common mode voltage isolation. Similar instrumentation amplifiers are available for powering isolated

Common-Mode Input Voltage

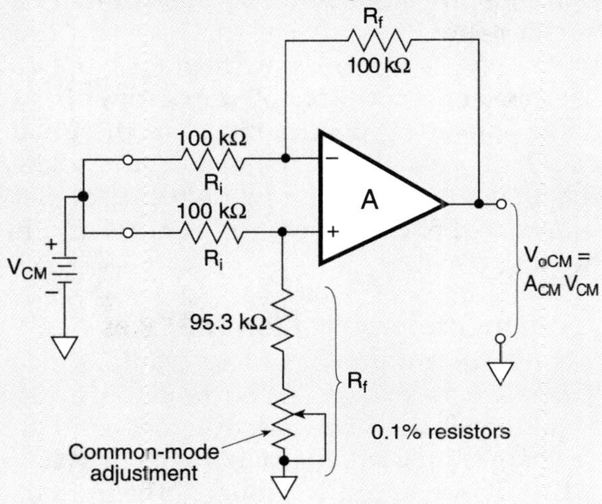

Fig. 10.07. Common-mode voltage is measured between the two input terminals and the common terminal. Because the two inputs must have identical voltages, they may be tied together and connected to one voltage source.

bridges, cold junction compensation, linearization, and other special signal-conditioning requirements. (See Figure 10.08.)

Optical Isolation

Optical isolation is now the most commonly used method to couple digital signals. The measured input voltage signal is converted to a current, which activates a light-emitting diode within an optical coupler. A light-sensitive transistor located adjacent to the diode, but on the opposite side of a voltage barrier, converts the light signal back to a current that the instrumentation amplifier can handle. The voltage barrier typically provides as much as several thousand volts of isolation between input and output.

Optical devices also are commonly used to isolate the output of an ADC, which is usually a serial string of data pulses passing through a single optical coupler. (See Figure 10.09.) The serial string is often converted from several parallel signals (ranging from 8 to 24 output ports, for example) to minimize the number of optical devices required in a system. Parallel to

Transformer Isolation Technique

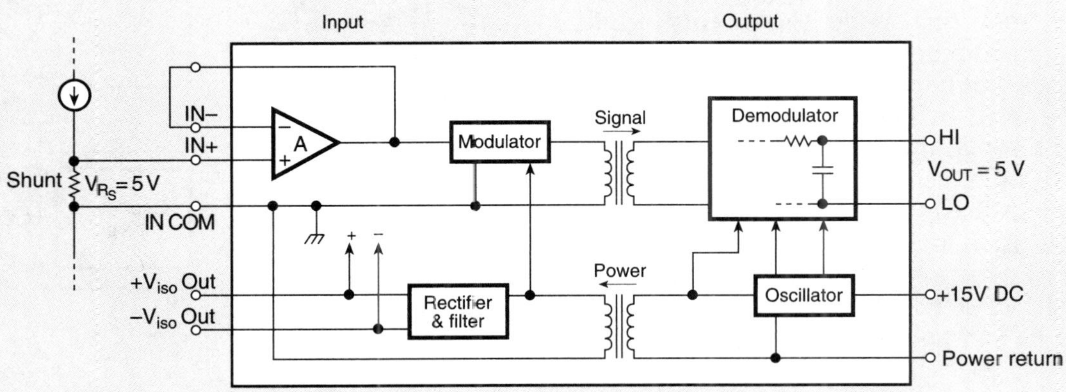

Fig. 10.08. This instrumentation amplifier with transformer coupling is typically used in an industrial process control loop and provides galvanic isolation in both input and output circuits. It can measure a ±5 Vdc signal riding on a common-mode voltage as high as 2000 V.

Optical Isolation Technique

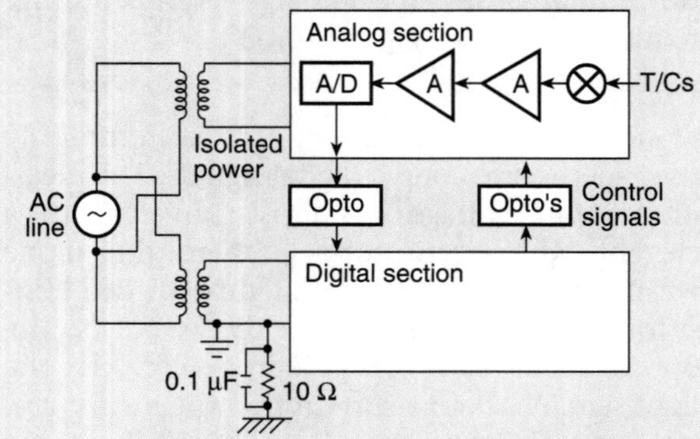

Fig. 10.09. Optical devices are the most widely used component for coupling signals across high-potential barriers and for low-level signals that are prone to the electrical noise typically coupled through different ground potentials.

serial conversion circuits are less expensive than the 8 to 24 optical devices (one for each bit of output from the parallel ADC). In these instances, the power supply for the ADC and associated input circuitry are also isolated, usually with a transformer.

Capacitive Isolation

A capacitor is a passive device that couples ac voltage from one stage to another while blocking the dc component. By this definition, it's a simple but inexpensive isolator. The measured signal to be isolated is modulated and coupled through the capacitor to the receiving side. On the receiving side, the ac signal is demodulated to restore the original signal. This technique is often applied to low-cost isolation amplifiers where the coupling capacitor is composed of a common layer between two isolated IC substrate sections. Signal isolation using these specialized ICs is rated as high as 1,500 V. The main benefits of this approach are simplicity, low cost, and bandwidths as high as 50 kHz. Figure 10.10 illustrates a dc/dc converter often used as the modulator/demodulator in an isolation amplifier.

Figure 10.11 illustrates a typical multiple-channel, programmable isolation amplifier in a data acquisition system that uses all three types of isolation: transformers, optical devices, and capacitors. A transformer-based dc/dc converter supplies power to the isolated side. A capacitively coupled device isolates the analog signal while two optical couplers transmit the digital control signals to the floating circuitry.

Fundamental Application Mistakes

Dangers in measurement can easily surface when a non-isolated analog input to a data acquisition system is erroneously connected to a device operating at high common-mode voltage with respect to earth ground. The mistake, which often precedes this problem, is determining the voltage range of the desired signal with a handheld digital or analog multimeter, and neglecting the relationship of the signal

Modulator/Demodulator Coupling

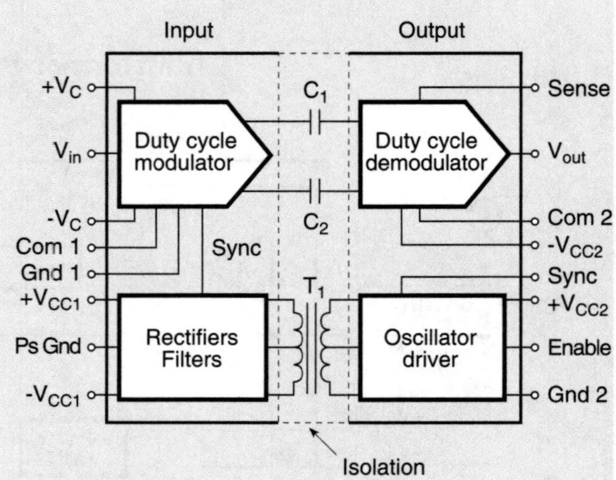

Fig. 10.10. Low-cost isolation amplifiers couple dc signals by modulating the signal, coupling its ac equivalent through a substrate capacitor connecting the stages, and then demodulating the signal to restore it to its original dc value.

Programmable Amplifier

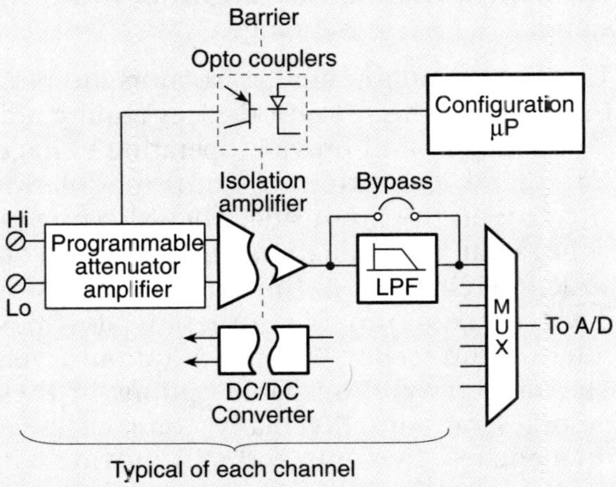

Fig. 10.11 All three techniques of isolation are used in a multi-channel, programmable amplifier. Optical devices couple digital signals, capacitors couple analog signals, and the dc/dc converter supplies power to the isolated side.

Isolation in Motor Drives

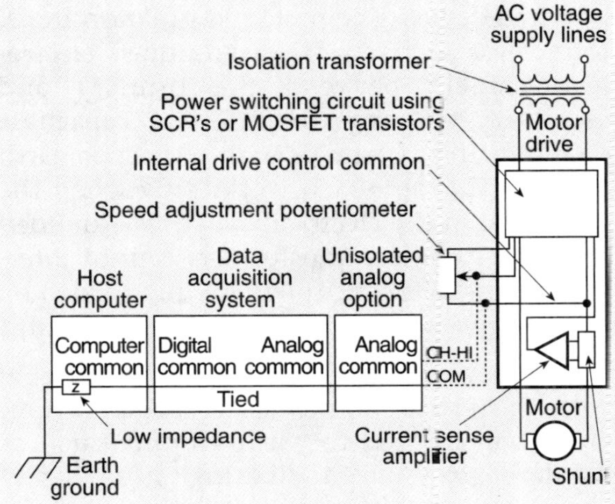

Fig. 10.12. Common-mode voltage is defined as the voltage between a signal and the localized common, which is usually earth ground. Common-mode voltages can be quite high, depending on the signal source. For example, a current sensing shunt in a dc motor drive might be expected to be at line-voltage potential, although the actual signal level is less than 50 mV across the shunt terminals. Measurement techniques for small signals at high common mode potentials call for isolation.

with respect to earth ground. If connected with the dotted lines, as shown in Figure 10.12, the motor drive can be damaged as soon as power is applied. This is because the control circuitry in many ac and dc motor drives is necessarily referenced to high voltage with respect to earth ground. When motor current is sensed with a shunt or low-value resistor, the control common bus is generally at high common-mode potential with respect to earth ground.

Some drives use non-contacting current sensors, transformers, and optical couplers to achieve galvanic isolation of the control circuitry. However, unless the drive is specifically known to have an isolated analog interface, assume it does not.

Isolation Transformers

Not all ac line-powered devices contain the internal isolation transformers that step the voltage down to lower operating levels needed for electronic circuits and simultaneously protect users from external ground faults. The power supply common bus in transformerless devices often connects to one side of the ac line cord. If the system is not protected from reversing the line cord connector in the socket, the device common line (and hence the enclosure) can be raised to a voltage level which is higher than the common terminal of other devices in the vicinity or other instruments connected to the data acquisition system. The ground fault resulting from this arrangement effectively shorts out the supply line and can be a hazard to operators and destroy the equipment.

An isolation transformer lets a user safely connect a data acquisition system input channel, grounded through the host computer, to a low voltage signal in an ac-powered device. The preferred isolation transformer for such use has a grounded electrostatic shield between the primary and secondary windings to minimize capacitive coupling and potentials with respect to earth ground. This approach works only when the ac neutral in the electrical system is grounded to earth. Isolation transformers cannot interrupt the path for the safety ground carried by the ground prong of a standard 3-wire cord.

A laptop computer running on an internal battery and not connected to other peripherals that are tied to earth ground (such as printers) can be a floating host safely connected to a data acquisition system. But a better overall approach is to isolate the input signal source.

An isolation transformer usually doesn't supply all the isolation intended for a data acquisition system, because the digital common in most computers provides a low impedance path to earth ground as part of its ESD (electrostatic discharge) protection scheme. But some data acquisition devices that communicate with the host computer through serial data links such as RS232 and RS485 use communication isolators specifically designed for that protocol. By comparison, most Ethernet interfaces are transformer isolated and have ESD protection networks on both sides of the isolation barrier referenced to chassis, plus earth ground through the host computer.

Analog Isolators

The ideal solution for measuring high common-mode signals is the analog isolator. The isolator safely measures low-level analog input signals containing as much as 1,500 V common mode, through magnetic, optical, or capacitive devices. The amplifiers provide both channel-to-system isolation and channel-to-channel isolation. By contrast, most solid-state multiplexers have no channel-to-channel isolation beyond the standard ±10-V signal range.

The most common analog isolators are plug-in modules. These 3-port devices require a dc power supply and provide operating voltages for signal conditioning and modulation circuitry on the input side. They also provide voltage for demodulation and signal reconstruction circuits on the output side. Most devices also provide inherent low-pass filtering and scaling to 0 to 5 V output levels. The wide array of available options in these modules can simplify many complex measurement requirements and still provide data to the overall data acquisition system of choice, as all data acquisition manufacturers have products that accept these modules.

The isolation modules are relatively expensive and are not likely to be used in low-cost data acquisition systems. Low-cost systems typically don't contain analog isolators, but many applications require isolators in at least a few channels. System-wise, the best place to address a high common-mode signal channel is at the source for safety and signal integrity.

Wireless Techniques

Not all data acquisition systems can connect to sensors on the test specimen with wires. They require a form of radio communication called telemetry. Radio transmitters and sensors are located on the device under test and receivers are located at the data acquisition system. For example, the rotating member of a large motor or generator can be monitored remotely and safely. The system can monitor temperature, vibration, deflection, and speed in rpm without the type of slip rings used in the past.

A relatively new protocol called Bluetooth is increasingly being used for remote measurement and control. It is a short-range wireless system that lets devices recognize, connect, and transfer data among them. The devices are equipped with special Bluetooth chips and transmit over a short range, typically 10 m. They can transfer data at a rate of 720 kb/sec over a frequency band of 2.40 to 2.48 GHz. Another system is the 802.11 Ethernet-based wireless system. It often provides physical and electrical isolation on the factory floor and for high voltage utility lines and demolition test sites. It operates in the same frequency range as Bluetooth and can handle higher data transfer rates of 1 to 11 Mb/sec.

NOISE REDUCTION

Signal Averaging

Some noise reduction techniques prevent the noise from entering the system initially and others remove extraneous noise from the signal. Another technique averages several signal samples through software. Depending on the nature of the noise and the specific averaging method, noise can be reduced by the square root of the number of averaged samples (RMS). But this may require abundant samples to obtain an acceptable measurement. Figure 10.13 shows the voltage across the shorted channel when only 16 samples of data are averaged.

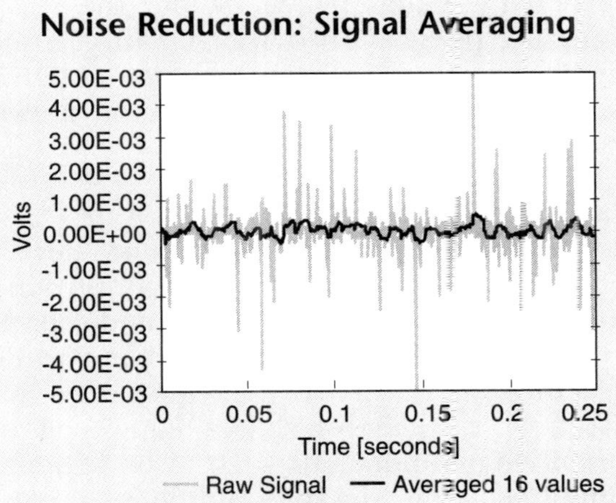

Fig. 10.13. Signal-averaging using software techniques virtually reduce the effective electrical noise pick up before the data acquisition system processes the measured signal.

Although averaging is an effective technique, it has several drawbacks. The noise present in a measurement sequence decreases as the square root of the number of measurements. Therefore, in the above example, reducing the RMS noise to a single count by averaging alone would require 3,500 samples. As such, averaging suits only low-speed applications, and it eliminates only random noise. It does not necessarily eliminate many other types of annoying system noise, such as periodic noise from switching power supplies.

Analog Filtering

A filter is an analog circuit element that selectively attenuates a particular band of frequencies in an incoming signal. Filter circuits can be passive or active. Depending on whether the filter is low or high-pass, it determines the frequencies that are attenuated above or below the cutoff frequency. For example, as a signal frequency increases beyond the cutoff point of a single-pole, low-pass filter, its attenuation increases slowly. Multiple-pole filter attenuation also increases slowly. Multiple-pole filters provide greater attenuation

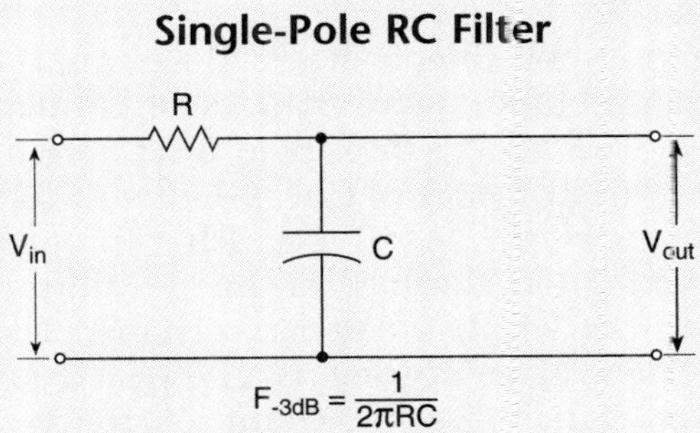

Fig. 10.14. Passive single pole filters are moderately effective in passing a band of frequencies that contain the measured signal, while attenuating other frequencies containing noise.

beyond the cutoff frequency, but they may introduce phase shifts that could affect some applications. The frequency where the signal is 3 dB down is given by the equation shown in Figure 10.14.

Passive vs. Active Filters

A passive filter is a circuit or device consisting entirely of non-amplifying components, typically inductors and capacitors, which pass one frequency band while rejecting others. An active filter, on the other hand, is a circuit or device composed of amplifying components such as operational amplifiers, and suitable tuning elements, typically resistors and capacitors, which pass one frequency band while rejecting others. Figure 10.15 compares the amplitude of a single-pole, low-pass filter with a three-pole filter. Both types are set for a 1-kHz cutoff frequency. The three-pole filter has a much greater attenuation for frequencies exceeding the cutoff. The improvement in signal quality provided by low-pass filtering is demonstrated in Figure 10.16 in which a signal containing wideband noise passes through a three-pole filter with a 1-kHz cutoff frequency. The deviation from the average signal is plotted in volts. The maximum deviation is 6 counts, and the RMS noise is 2.1 counts.

The three-pole filter shown in the example has an active input with changeable configurations. The active, three-pole filter can be a Butterworth, Bessel, or Chebyshev with corner frequencies up to 50 Hz. Filter properties depend on the values of the resistors and capacitors, which the user can change. Filters also use switched-capacitors. This type requires a clock signal to set the cutoff frequency. The primary advantage of this filter is the ease of programming the cutoff frequency.

Differential Voltage Measurement

Differential input amplifiers are most often

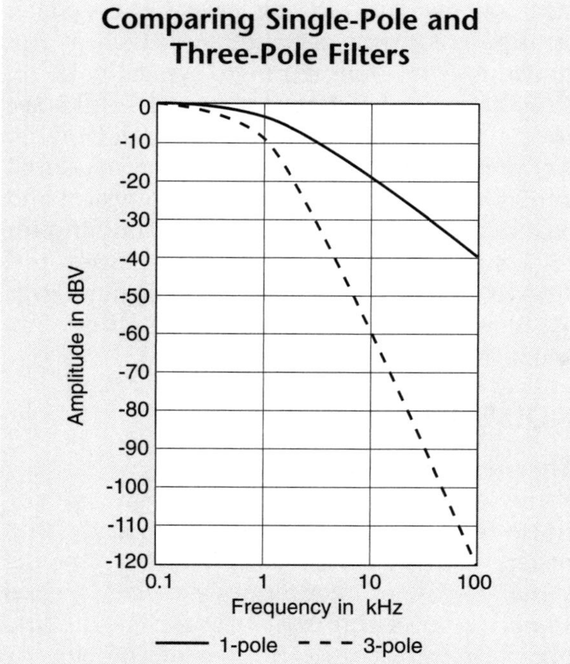

Fig. 10.15. The three-pole filter attenuates the frequencies after the cutoff point more effectively than does the single-pole filter.

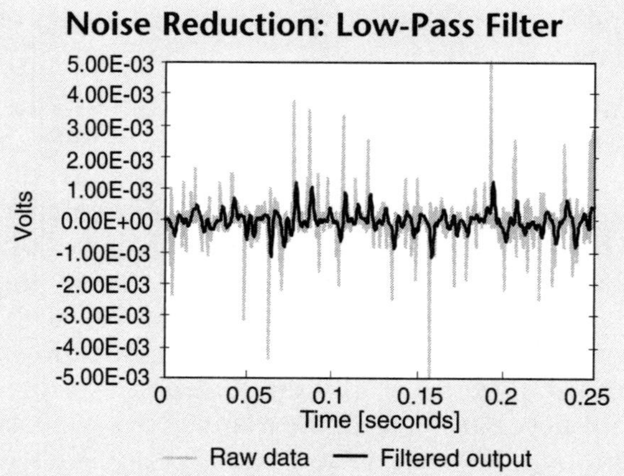

Fig. 10.16. An active, three-pole low-pass filter, configured as a Butterworth, Bessel, or Chebyshev filter, reduces the noise signal more effectively without significant attenuation than does a passive circuit.

used in data acquisition systems because they provide a high gain for the algebraic difference between their two input signals or voltages, but a low gain for the voltages common to both inputs. Making differential voltage measurements is another means of reducing noise in analog input signals. This technique is effective because often, most noise on the high-side input lead closely approximates the noise on the low lead. This is called common-mode noise. Measuring the voltage difference between the two leads eliminates this common-mode noise.

The improvement gained with differential voltage measurements is illustrated in Figure 10.17. It shows the same signal as Figure 10.05, but using a differential input rather than a single-ended input.

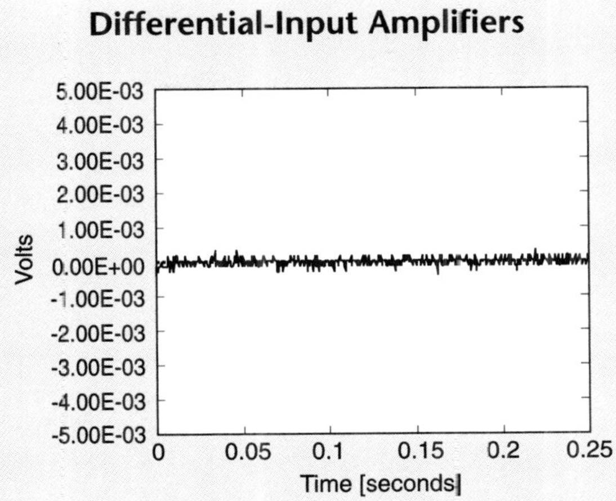

Fig. 10.17. *The biggest improvement in reducing unwanted noise signals from the measured variable comes from differential-input amplifiers. It works so well because most of the noise on the high side duplicates the noise on the low side of the input and the algebraic sum of the two equal parts.*

Chapter 11

DIGITAL AND PULSE-TRAIN CONDITIONING

DIGITAL I/O INTERFACING

Digital Signals

Digital signals are the most common mode of communications used between computers and peripherals, instruments, and other electronic equipment because they are, of course, fundamental to the computers' operation. Sooner or later, all signals destined to be computer inputs must be converted to a digital form for processing.

Digital signals moving through the system may be a single, serial stream of pulses entering or exiting one port, or numerous parallel lines where each line represents one bit in a multi-bit word of an alphanumeric character. The computers' digital output lines often control relays that switch signals or power delivered to other equipment. Similarly, digital input lines can represent the two states of a sensor or a switch, while a string of pulses can indicate the instantaneous position or velocity of another device. These inputs can come from relay contacts or solid-state devices.

High Current and Voltage Digital I/O

Relay contacts are intended to switch voltages and currents that are higher than the computers' internal output devices can handle, but the frequency response of their coils and moving contacts is limited to relatively slowly changing I/O signals or states. Also, when an inductive load circuit opens, its collapsing magnetic field generates a high voltage across the switch contacts that must be suppressed. A diode across the load provides a path for the current spike while the inductor's magnetic field is collapsing. Without the diode, arcing at the relay's contacts can decrease its life. (See Figure 11.01.)

Flyback Diode Protection

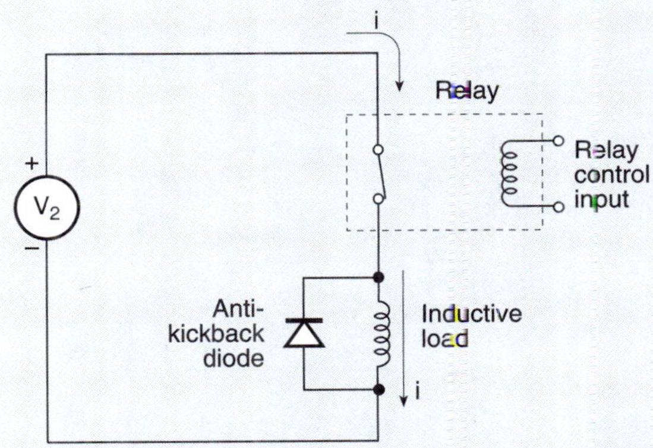

Fig. 11.01. *The fly-back diode clips high voltage spikes ordinarily developed across the inductive load when the control relay contacts open. Without diodes, the high voltage arcs across the opening contacts, substantially reducing their useful life.*

TTL and CMOS devices usually connect directly to high-speed, low-level signals, such as those used in velocity and position sensors. But in applications where the computer energizes a relay coil, TTL or CMOS devices may not be able to provide the needed current and voltage. So a buffer stage is inserted between the TTL signal and the relay coil, typically to supply 30 V at 100 mA.

An example of this type of system is an optional card for a digital I/O instrument. It contains an amplifier/attenuator stage, consisting of a PNP transistor, a fly-back diode, and a resistor. (See Figure 11.02.) To energize a standard 24-V relay, an external 24-V supply is connected to the circuit. As the internal TTL output goes high, the transistor is biased and the output goes low, (about 0.7 V). When the TTL output is low, the transistor stops conducting and the output goes to 24 V. Since the relay coil is an inductive load, the fly-back diode should be attached to prevent damage during switching.

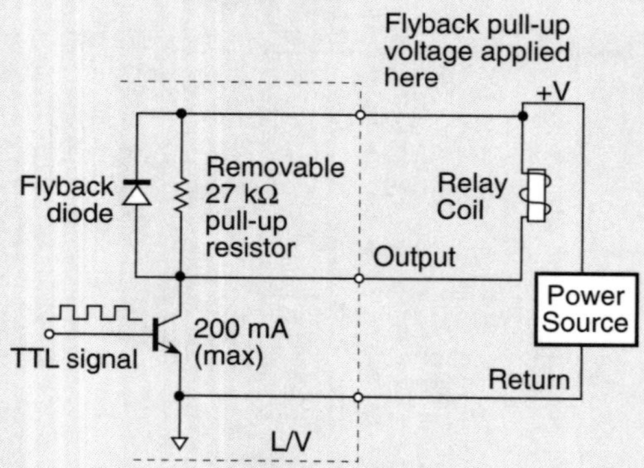

Fig. 11.02. A plug-in card containing a special digital amplifier is often provided to interface a TTL signal to a control relay. The amplifier can boost the standard 5-V TTL signal to 24 Vdc at 200 mA.

Figure 11.03 shows a high-voltage digital input with an attenuator circuit. This allows the TTL circuitry to read voltages up to 48 V. The high-voltage signal connects to a resistive voltage divider, which is a signal attenuator. Selecting an appropriate resistance value R provides a means for selecting the high-voltage level. The table in Figure 11.04 shows the resistor values for frequently used levels.

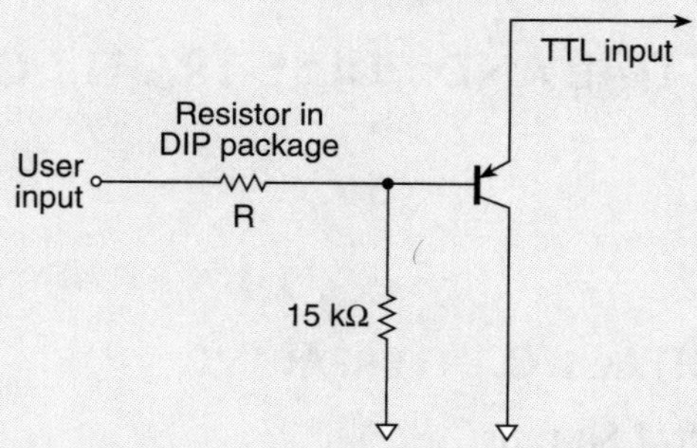

Fig. 11.03. One optional interface card comes with an adjustable attenuator to handle an input range from 5 to 48 V in four steps. The table specifies the voltage levels that can be selected with four different resistors.

Typical Voltage Levels & Associated Resistor Values

Inputs	Resistor Value, R
0 to 5 V	10 Ω
0 to 12 V	20 kΩ
0 to 24 V	56 kΩ
0 to 48 V	120 kΩ

Fig. 11.04. Typical voltage levels and resistor values.

DIGITAL INPUTS

The methods used for interfacing digital inputs to a computer range from simple to complex. This section briefly discusses software-triggered, single-byte readings; hardware-paced, digital input readings; and externally triggered, digital-input readings.

Asynchronous Digital-Input Readings

A software-triggered, asynchronous reading is required when the computer periodically samples a digital byte or a group of bits. And sometimes the speed and timing of the digital-input readings are particularly critical. But the time between readings is likely to vary when using the software-triggered, single-byte method, particularly in applications running under a multitasking operating system such as in a PC. The reason is that the time between readings depends on the speed of the computer and other tasks that must be performed concurrently. Variations in time between readings can be partially compensated with software timers, but a timing resolution of less than 10 ms is not guaranteed on a PC.

Synchronous Digital-Input Readings

Some systems offer hardware-paced digital input readings. In such systems, the user sets the frequency that the digital input port can read. For example, one system can read its 16-bit port at 100 kHz, while another operates to 1 MHz. The greatest advantage of hardware-based digital input readings is that they can be implemented far more quickly than can software-triggered readings. Finally, devices such as these can locate the digital-input port readings among the analog readings, providing a close correlation between analog and digital input data.

Externally Triggered Digital Input Readings

Some external devices provide a digital bit, byte, or word at a rate independent of the data acquisition system. They take readings only when new data are available rather than at a predetermined interval. Because of this, such external devices typically transfer data through a handshaking technique. As new digital information becomes available, the external device issues a digital transition on a separate line, such as an External Data Ready or Strobe input. To interface with a device like this, the data acquisition system must provide an input latch, which the external signal controls. Furthermore, a logic signal supplied to the controlling computer alerts it to the fact that new data are ready to be received from the latch.

One example of a device that operates in this fashion has an inhibit line among its six handshake/control lines for notifying external devices that the input latch is being read. This procedure lets the external device hold new digital information until the current read event is performed successfully.

DIGITAL ISOLATION

Digital signals often are isolated for several reasons: To protect each side of the system from an inadvertent over-voltage condition on the opposite side, to facilitate communication between devices with different grounds, and to prevent injury when circuits are attached to people in medical applications. One common approach for isolation is through an optical coupler. Optical coupling consists of an LED or diode laser to transmit the digital signal and a photodiode or phototransistor to receive it. (See Figure 11.05.) Small optical couplers isolate voltages up to 500 V. For example, this technique effectively controls and monitors digital devices connected between dissimilar grounds.

PULSE TRAIN SIGNAL CONDITIONING

In many frequency-measurement applications, pulses are counted and compared against a fixed time base. A pulse can be considered a digital signal because only the number of rising or falling edges is measured. In many instances, however, the pulse-train signal comes from an analog source, such as a magnetic pickup.

For example, one widely used input frequency card in a data acquisition system provides four channels of frequency input through two separate front-end circuits, one for true digital input circuits and one for analog inputs. The card conditions digital inputs of different levels, and the analog input circuit converts a time-varying signal into a clean digital pulse train.

Optical Coupler Isolation

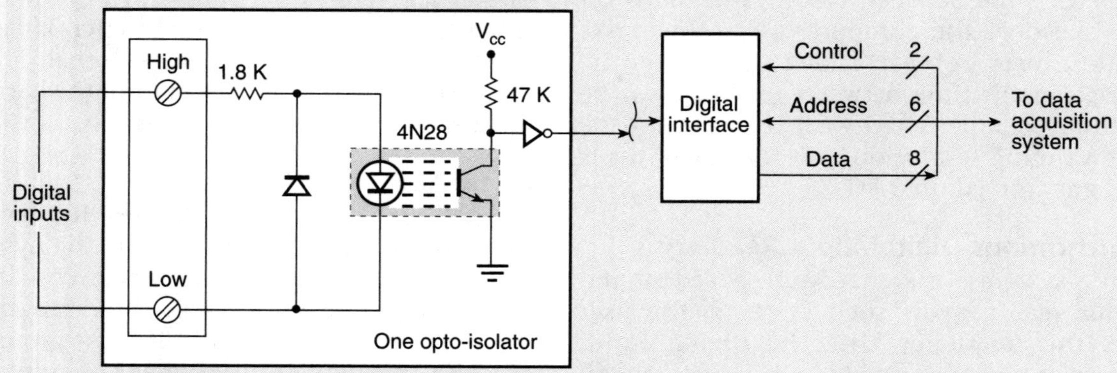

Fig. 11.05. An optical isolator decouples the digital signal source ground from the data acquisition system ground, and protects the acquisition system inputs from unintentional high voltages.

Fig. 11.06 shows the schematic of the analog input, signal-conditioning path. The front-end RC network provides ac coupling allowing all signals above about 25 Hz to pass. The selectable attenuator reduces the overall magnitude of the waveform to desensitize the circuit from unwanted low-level noise. When using a pulse train from a relay closure, the unit provides programmable settings that let the user select the amount of debounce time required. The digital circuitry monitors the conditioned pulse-train for a sustained high or low level. Without debouncing, the extra edges in the signal produce an excessively high and erratic frequency reading. (See Fig. 11.07.)

Many transducers generate frequency-modulated output signals rather than amplitude-modulated. For instance, sensors that measure rotational motion and fluid flow typically fall into this class. Photomultiplier tubes and charged-particle detectors also are often used for measurements that require pulse counting. In principle, such signals could be sampled with an ADC, but this approach generates much more data than necessary and makes the analysis cumbersome. Direct frequency measurements are far more efficient.

FREQUENCY-TO-VOLTAGE CONVERSION

Data acquisition systems measure frequency in several ways; they integrate a continuous wave ac signal or pulse trains to produce a dc voltage with a magnitude proportional to the frequency, convert the ac voltage to a binary digital signal with an ADC, or count digital pulses.

Pulse-Train Integration

One conversion technique commonly used in a single-channel, modular signal conditioner

Analog to Pulse Converter

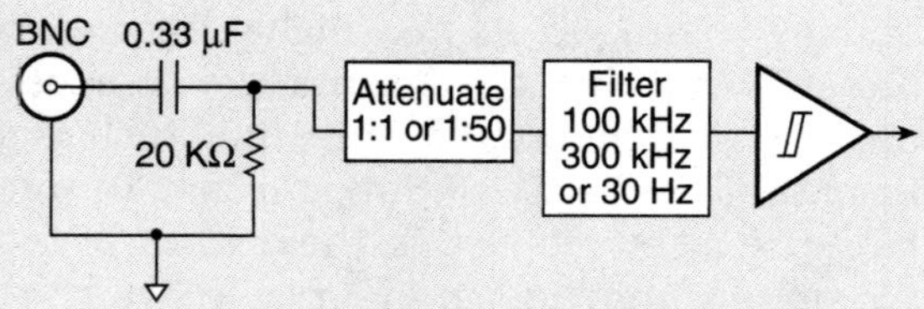

Fig. 11.06. An interface circuit contains an RC network that significantly attenuates ac signals below 25 Hz before converting the higher frequencies to a digital pulse stream.

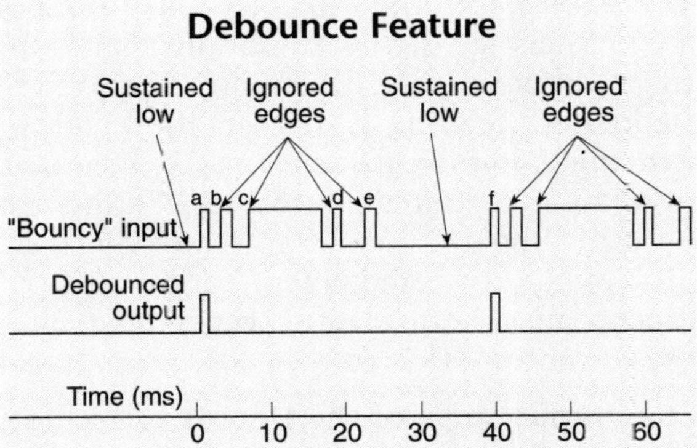

Fig. 11.07. A relay contact debounce circuit filters extraneous signals that the measurement amplifier could mistake for rising and falling edges of a legitimate string of signal pulses.

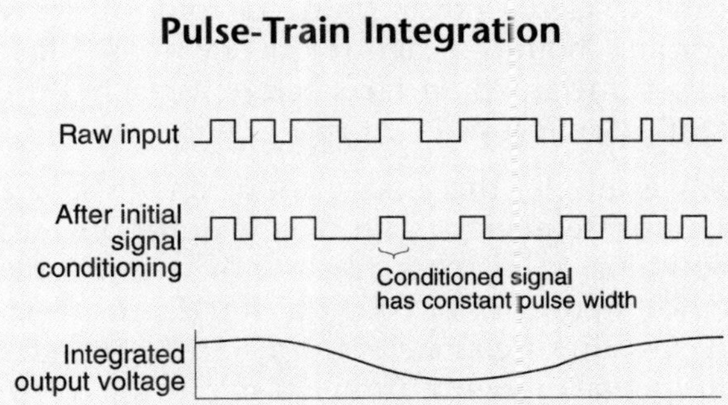

Fig. 11.08. In one type of frequency-to-voltage converter, an integrator circuit converts a string of constant-width pulses of varying repetition rate into a smoothly changing analog voltage. The analog signal is proportional to the input frequency.

integrates the input pulses and produces an output voltage proportional to the frequency. First, a series capacitor couples the ac signal, which removes extremely low frequency ac and dc components. A comparator generates a constant pulse width each time the input signal passes through zero. The pulse then passes through an integrating circuit such as a low-pass filter and generates a slowly changing signal level at its output, proportional to the input frequency. (See Fig. 11.08.)

The response time of the frequency-to-voltage converter is low – the inverse of the cutoff frequency of the low-pass filter. This cutoff frequency should be much lower than the input frequencies being measured, but high enough to provide the required response time. As the measured frequency approaches the cutoff frequency however, significant ripple in the output becomes a problem as shown in Figure 11.09.

An external capacitor selects the time constant for an IC dedicated to frequency to voltage conversion. The circuit can measure signals in vastly different fixed frequency ranges, but the capacitor must be changed to change a frequency range. Unfortunately, such frequency-to-voltage converters work relatively poorly for frequencies below 100 Hz because a low-pass filter with a cutoff frequency under 10 Hz requires an excessively large capacitor.

Digital-Pulse Counting

Another type of conversion technique measures the frequency of a string of digital pulses or an ac-coupled analog signal voltage. It outputs a dc voltage level proportional to the input frequency, similar to the integrator described above. However, the dc level here comes from a DAC output. Front-end circuitry converts the incoming analog or digital signal

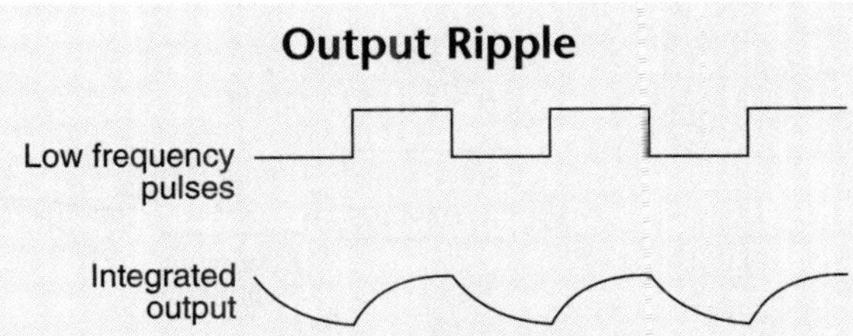

Fig. 11.09. The pulse-train integrator handles only frequencies above a certain cutoff, due to the integrator's time constant. Frequencies below cutoff produce output ripple if allowed to pass.

into a clean pulse train, devoid of relay contact bounce, high-frequency noise, and other unwanted artifacts before it reaches the DAC. (See Figure 11.10.)

For example, the analog input channel of a typical frequency-input data acquisition card contains a low-pass filter with a selectable cutoff frequency of 100 kHz, 300 Hz, or 30 Hz. It measures frequencies from 1 Hz to 100 kHz for signals ranging from 50 mV to 80 Vp-p. The digital input circuit measures ±15-Vdc signals from 0.001 Hz to 950 kHz, dc-coupled to a TTL Schmidt trigger circuit. The cards typically come with pull-up resistors for use with relays or switches.

A microcontroller accurately measures a total period consisting of several cycles extending over one user-selectable minimal period, which determines the frequency resolution. The microcontroller computes the frequency from the measured period and converts it to a command for a DAC, which, in turn, provides the dc level to the data acquisition system. The dc output of the DAC drives the input of an ordinary dc signal conditioner, and the software converts the dc level to an equivalent frequency reading. This method allows extremely low frequencies to be measured over an exceptionally wide range, and the output update can be relatively fast. Moreover, the frequency range can be programmed, letting the expected frequencies use the entire ADC range.

The output range of the DAC is +5 to –5 V. The minimum frequency selected by the user becomes the –5 V output while the maximum frequency becomes +5 V. Virtually any frequency bandwidth may be selected such as 500 Hz F_{min} to 10 kHz F_{max}, or 59.5 Hz F_{min} to 60.5 Hz F_{max}. With an ADC of 12-bit resolution, the lower bandwidths will have higher resolution than higher bandwidths, simply because the range from –5 to +5 Vdc is partitioned into 4096 parts regardless of bandwidth. The one-Hz bandwidth is divided into 4096 parts, which yields a resolution

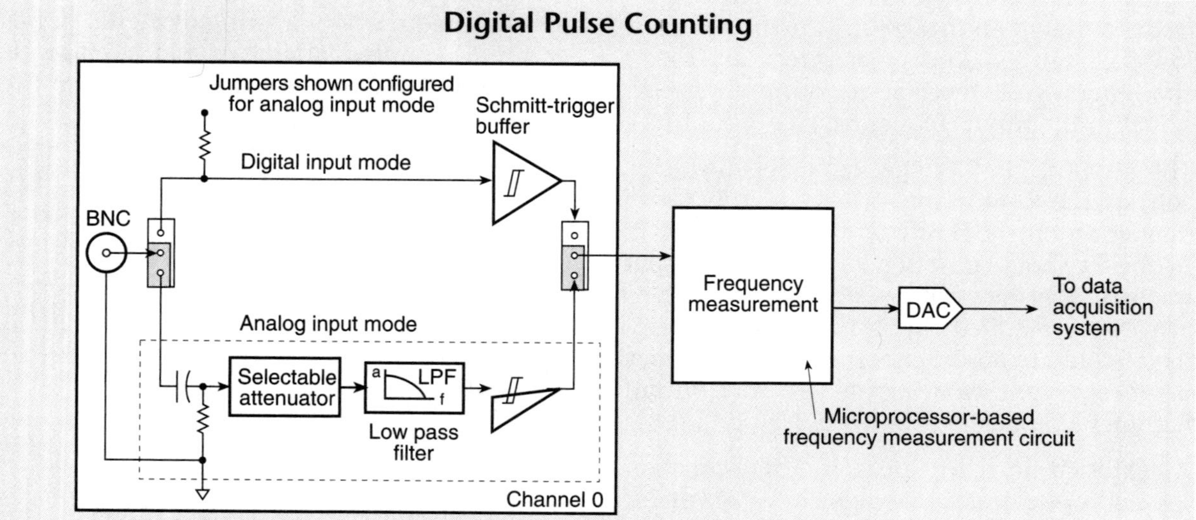

Fig. 11.10. In the other type of frequency-to-voltage converter, the input card contains an input coupling capacitor to filter low frequencies, an attenuator, a low pass filter, and a microprocessor to calculate the frequency. The signal is then connected to a DAC before feeding into the data acquisition system.

of 1/4096 Hz or about 0.00244 Hz. For the 100 kHz bandwidth, the resolution becomes 24.41 Hz.

The resolution is 12 bits over all ranges, but the update time depends on the range selected. From 1 Hz to the user-defined maximum upper range boundary, the voltage conversion update is 2 to 4 ms or the period of the input frequency, whichever is greater. For a range of 0 to 10 kHz, the update rate is 2 to 4 ms, and for a range of 0 to 60 Hz, the output updates every cycle or 16.6 ms. As the conversion range becomes narrower, from 49 to 51 Hz, for example, the time to resolve the 2-Hz differential to 12-bit resolution increases. In this case, the conversion time is approximately 59 ms.

In addition to the low-pass filter, a predefined hysteresis level is built in to help prevent false counting caused by high-frequency noise. A debounce time can be programmed from 0.6 ms to 10 ms for handling electromechanical devices such as switch or relay contacts that bounce or chatter while switching.

Frequency Measurement by Gated-Pulse Counting

Gated-pulse counting can measure frequencies much more accurately than frequency-to-voltage conversion methods. Gated-pulse counting methods count the pulses that appear over a specified period of time. Dividing the number of pulses by the counting interval determines the frequency, and the error can be as low as the inverse of the counting interval. For example, if the counting interval is two seconds, the error can be as low as 0.5 Hz.

Many data acquisition systems include TTL-compatible counter/timer ICs that can perform gated-pulse, digital-level input, however, they are unsuitable for unconditioned analog signals. Fortunately, many frequency-output devices feature a TTL output option. Some products use a counter/timer IC, which contains five counter/timers. Many counter/timer ICs generally use an oscillator built into the data acquisition system, or an external oscillator. Such ICs usually have several channels available to assist counting applications. Each channel contains an input, a gate, and an output. The simplest counting method only uses the input, and the PC is programmed to periodically read and reset the counter. The weakness in this approach is the uncertainty in the timing interval. Variations crop up in the execution speeds of the functions that begin and end counting. In addition, the function call that delays execution of the program for 50 ms runs under an inaccurate software timer. These two effects can render a short-counting interval, frequency measurement useless. However, the technique is usually sufficient for counting intervals greater than one second.

Gating can attain greater accuracy because the gate controls the counting interval. Consequently, frequency measurements are independent of any software timing issues. The gate can be configured so that pulses are counted only when a high-level signal enters it. Similarly, the gate can start counting when it detects one pulse and stop counting when it detects another.

A disadvantage of gated-pulse counting is that it requires an extra counter to provide the gate. However, in multiple channel applications, a single counter can provide the gate for many channels. For example, in a five-channel system, four channels count while one channel provides a gate.

Timing Applications

A counter/timer also can be used in a data acquisition system for timing applications. A clock signal connected to the input of a channel and using the input signal as a gate works well. The method requires that the gate be configured for counting when the gate input is high. A similar technique can measure the length of time elapsed between two pulses by configuring the gate to begin counting at the first pulse and end counting at the second.

Because a 16-bit counter overflows at 65,535 counts, the maximum pulse width measurable with a 1-MHz clock is 65,535 ms; a longer pulse overflows the counter. However, a clock slower than 1-MHz can be used for longer pulses.

Chapter 12

TEDS: TRANSDUCER ELECTRONIC DATA SHEETS

SMART SENSORS

Certain types of transducers are classified as smart sensors. They contain a Smart Transducer Interface Module (STIM), which in turn, contains an onboard EEPROM memory IC, called TEDS, Transducer Electronic Data Sheet. (See Figure 12.01.) The TEDS electronically stores information regarding the transducer's characteristics and parameters such as type of device, manufacturer, model number, serial number, calibration date, sensitivity, reference frequency, and other data. The TEDS also stores the calibration coefficients and frequency response for a transducer in terms of a table or an algorithm.

TEDS: Principal Parts

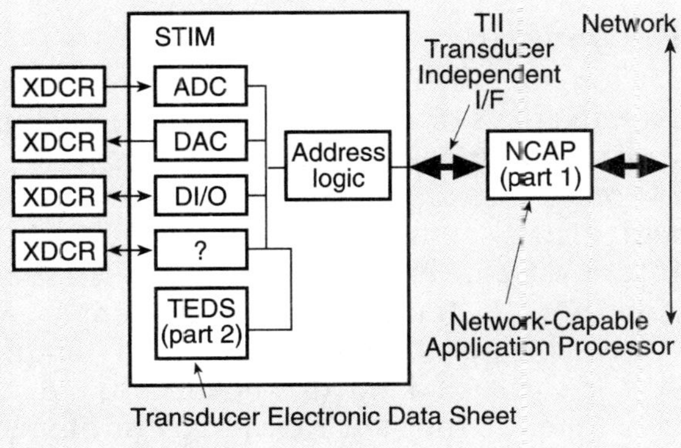

Fig. 12.01. Three principal parts comprise smart sensor systems: the Smart Transducer Interface Module STIM, the Transducer Electronic Data Sheet TEDS, and the Network-Capable Application Processor NCAP.

Not all sensor manufacturers provide this feature, but those that do are encouraged to follow an evolving standard, IEEE 1451.0. One part of the standard, 1451.4, defines the TEDS format, channel identification formats, electrical interface, and read and write functions for accessing the TEDS and transducer data. The specification also defines the data set, that is, the number of samples acquired for one command that varies from 0 to 65,535 samples per set.

The standard does not specify requirements for signal conditioning, signal conversion, or how applications can use the TEDS data. However, the signal conditioners and other interface hardware used with these smart sensors must provide for an option or include circuits for TEDS communications under the direction of a software module specifically intended for this purpose. Another special module that works with TEDS is usually part of the data acquisition system's application software package that collects, stores, and displays the measured variables. The hardware automatically scales the readings and sets the range according to the data stored in the sensor. This is typically done for each TEDS-associated channel.

TEDS capability was originally intended for piezoelectric sensors such as accelerometers and pressure sensors, but it now includes all common analog sensors and actuators, such as MEMS (micro-electromechanical sensors), accelerometers, pressure transducers, and temperature sensors with two-wire and mixed-mode (analog and digital) input/output.

For two-wire analog sensors and actuators (called Class 1), the output signal is generally coupled to the signal conditioner or driver while the TEDS data are enabled and read out with a dc bias voltage applied to the same two wires. (See Figure 12.02.) The transducers with mixed-mode capability (called Class 2) also communicate digitally with the TEDS memory. (See Figure 12.03.) The analog part is the signal representing the measured variable, and the digital interface communicates with the embedded EEPROM. The TEDS architecture contains standard templates for common transducers, and custom sub-templates that manufacturers can use for defining special parameters and custom requirements.

Class 1 Sensors: Analog

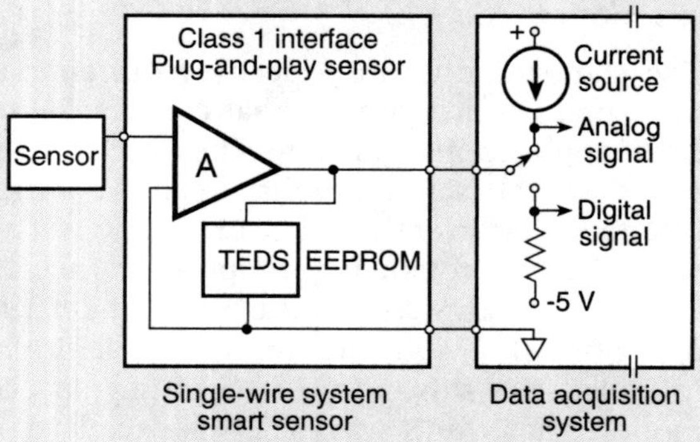

Fig. 12.02. Class 1 sensors have only two leads that connect to the input of the STIM. The analog signal voltage and any required biasing voltage share the same pair of leads.

The TEDS file may be contained onboard the sensor in the EEPROM, or off board in a reserved file in the data acquisition system. A number of transducer manufacturers plan to launch an Internet-accessible site where they list sensors that presently do not contain a TEDS memory IC. Users may enter the serial number of their sensor and obtain an equivalent TEDS specification to download. This lets application software programs work with both older sensors and newer TEDS-equipped sensors.

Class 2 Sensors: Mixed Mode

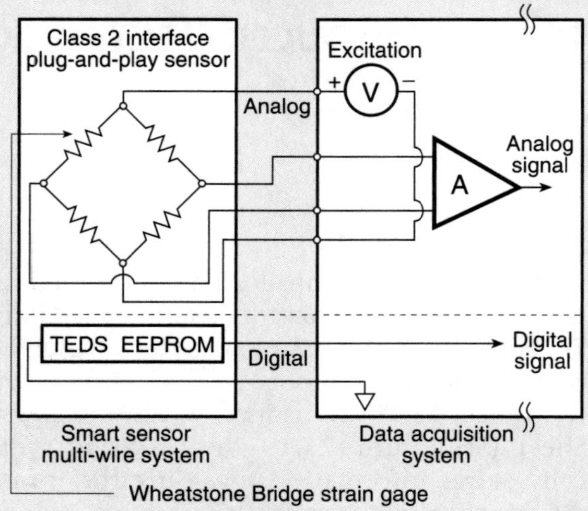

Fig. 12.03. Class 2 sensors use multiple leads to connect between the sensor and STIM. In the case of a Wheatstone Bridge circuit, for example, four leads connect the sensor and the other pair connects to the onboard TEDS EEPROM, which illustrates the mixed-mode application.

Networks

The original intent of the TEDS concept is to handle arrays of smart sensors and actuators over various types of networks. Different types of control networks now exist for both analog and digital communications, but most are intrinsically incompatible with one another. And not all transducer manufacturers are either able or willing to provide a unique sensor for each kind of network bus. As a consequence, they are increasingly turning to digital networks controlled with microprocessors to develop a universally accepted interface that is more economical for them to interconnect systems, networks, and instruments. The guiding document is the evolving

IEEE 1451 standard, which also allows the new devices to be compatible with older and existing systems.

System Architecture

The IEEE has partitioned the P1451 standard into five major parts. Each part addresses a different facet or interface of the project that enables sensor manufacturers to design and build new sensors compatible with all networks. Each part is defined as follows:

Part 1. *Network-Capable Application Processor (NCAP) Information Model.*
The purpose of this part is to define an object model for smart transducers that connect to a network, and specify a software interface to work with the transducer's components. The major components of this model include the NCAP block, function block, and transducer block. The NCAP processor connects the network to the transducer modules, and each different network requires a different NCAP physical interface. The smart transducer object model interfaces to the NCAP processor and to the transducer block. The interface to the transducer block contains details of the transducer hardware in a program model, and the interface to the NCAP block contains details describing the network protocols.

The NCAP's primary purpose is to communicate between the STIM and a particular network. The NCAP also computes calibration corrections and converts between values in metric units and values coming from the STIM's DACs or going to the ADCs.

Part 2. *Transducer to Microprocessor Communications Protocols and Transducer Electronic Data Sheet (TEDS) Format*
Part 2 defines the details of a TEDS, including the format for storing data in a small (256-bit to 4 kb) EEPROM and the interface between the NCAP processor and the transducers. The TEDS contains the information needed by the software to convert the sensor values such as volts or resistance to physical units, such as force in pounds or acceleration in gs. TEDS is also part of the Smart Transducer Interface Module. STIM includes the ADCs, DACs, digital I/O, and triggers that connect to the transducers.

Part 3. *Digital Communications and TEDS Formats for Distributed Multidrop Systems.*
This part is intended to define the TEDS format and the standard for the interface between multiple transducers in a multidrop network. Some of the issues dealing with a multidrop system include automatically identifying the transducer when it connects to the bus, and how quickly the system recovers after a short power dropout.

Part 4. *Mixed-Mode Communication Protocols and TEDS Formats*
Mixed-mode communications deals with the issue of using two-wire I/O, which shares both signal and digital interfaces vs. a multi-wire system where signal and digital communications are handled on separate wires or ports. Small analog sensors contain TEDS that let the sensors interface on the network through the same two wires they use for signals. The digital TEDS data uses the same two wires.

Part 5. *Wireless Communication Protocols and TEDS Formats*
At the time of this writing, the IEEE P1451 standards committee study group had defined the details of the interfaces for wireless communications, but not the modules. The intent of the standard is to separate the physical layers from the upper layers of the protocol stack. Significant progress on the specification is expected in 2004.

TEDS composition

The data structure for the TEDS works well with a variety of sensor types. Regardless of the sensor's principle of operation, the structure contains three major subdivisions;

TEDS Table	
Meta-TEDS	Provides operational parameters for the transducer
Transducer-channel TEDS	Provides operational parameters for the transducer
Calibration TEDS	Provides calibration coefficients for the transducer
Text-based TEDS	Provides a description of the device for the operator, human readable in XML
Frequency-response TEDS	Provides frequency response for a transducer and associated electronics in a table
Transfer-function TEDS	Provides frequency response for a transducer and associated electronics in algorithm form
End-user, application-specific TEDS	User writeable block of non-volatile memory. Contents defined by user.
Commissioning TEDS	User writeable block of non-volatile memory. Intended to provide an identifier for the transducer.
Manufacturer-defined TEDS	Format defined by the manufacturer
Physical TEDS	Defined for the particular physical layer, not included in P1451.0. How to access it is defined by IEEE P1451.0.

Fig. 12.04. Table 1. *Basic TEDS Table.*

the Basic TEDS; the Standard TEDS (subtemplate), and the User Area. (See Figure 12.04, Tables 1, 2, and 3.)

The Basic TEDS contains the manufacturer's identification, model identification, version letter, and the serial number, which are common to all types of sensors. The Standard TEDS, however, contains data unique to the sensor's principle of operation. For instance, a piezoelectric accelerometer differs from a strain gage load cell, so the Standard TEDS EEPROM contains some common and some unique information. Both sensor TEDS list Calibration Date, Measurement Range, and Electrical Output, but they differ in as many as 12 other parameters. The accelerometer has 6 parameters that the strain gage lacks, and the strain gage has 6 parameters the accelerometer lacks. The User Area contains information such as Sensor Location, Calibration Due Date, and Calibration Table.

TEDS Configuration: Accelerometer		
Basic TEDS	Manufacturer ID	25
	Model ID	5226
	Version Letter	A
	Serial Number	11521S
Standard TEDS	Calibration Date	Jan. 6, 2004
(Subtemplate)	Sensitivity @ Reference	1.12E+03 mV/g
	Reference Frequency	110.0 Hz
	Reference Temperature	25°C
	Measurement Range	±50 g
	Electrical Output	±5 V
	Quality Factor	
	Temperature Coefficient	-0.27%/°C
	Direction (X, Y, Z)	X
User Area	Sensor Location	Matrix A: B-3
	Calibration Due Date	Jun. 15, 2004

Fig. 12.04. Table 2. *Example for an Accelerometer.*

TEDS Configuration: Strain-Gage Load Cell		
Basic TEDS	Manufacturer ID	56
	Model ID	350A
	Version Letter	E
	Serial Number	150-00878T
Standard TEDS	Calibration Date	Mar. 15, 2004
	Measurement Range	±250 lb
	Electrical Output	±5 mV/V
	Bridge Impedance	500 Ω
	Measurement Range	±75 g
	Excitation, Nominal	24 Vdc
	Excitation, Minimum	10 Vdc
	Excitation, Maximum	
	Response Time	
Calibration TEDS	Calibration Table	-2.06 mV/V, -150 lb
		-1.02 mV/V, -75 lb
		0.03 mV/V, 0 lb
		0.95 mV/V, 75 lb
		1.89 mV/V, 150 lb
User Area	Sensor Location	L23-3
	Calibration Due Date	Jun. 15, 2004

Fig. 12.04. Table 3. *Example for a Strain-Gage Load Cell.*

Index